POWER BI FOR BEGINNERS

The 7-Day Method to Build Dashboards
Data Strategist (Video Course In

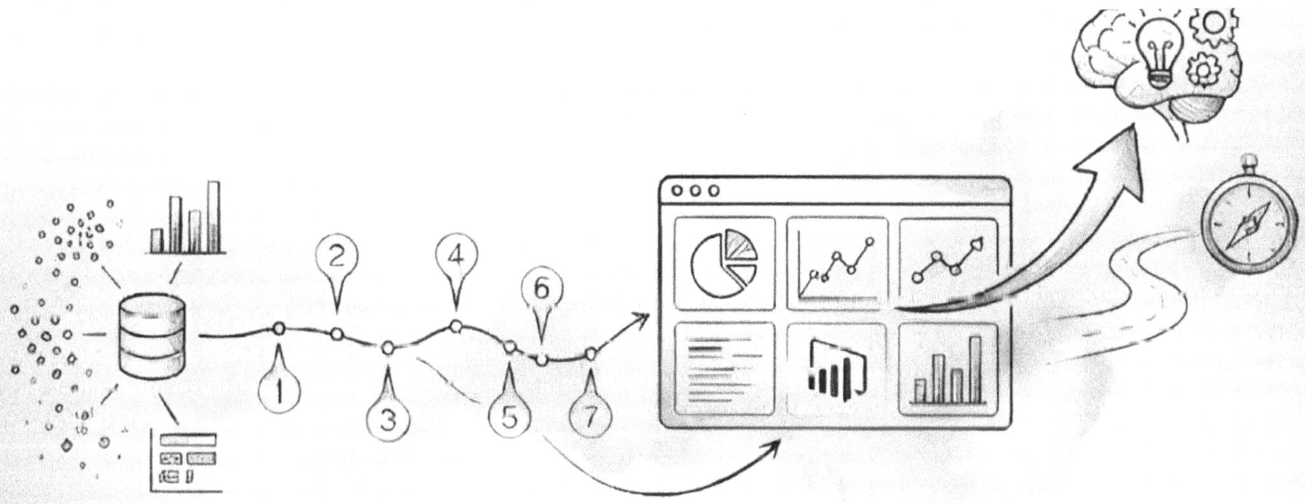

Kora Pierce Callum Pierce

Disclaimer

The contents of this book have been compiled and reviewed to the best of the authors' knowledge and with the greatest possible care. This information is intended for general informational purposes only and is not a substitute for professional, medical, legal, financial, or therapeutic advice.

The authors assume no responsibility for the timeliness, accuracy, completeness, or applicability of the content provided. Use of this information is at the reader's sole risk. The authors expressly disclaim any liability—regardless of legal grounds—for direct or indirect damages, particularly health-related, financial, or operational, that may result from the application of the information in this book, unless willful intent or gross negligence is proven.

Trademarks

All trademarks and product names mentioned are the property of their respective owners and are used for illustrative purposes only. No commercial or legal affiliation with these third parties exists.

TABLE OF CONTENTS

TABLE OF CONTENTS

Introduction: The Great Shift

The Excel Ceiling: When Spreadsheets Become a Liability

You know the specific, sinking feeling I'm talking about. It usually strikes at 5:30 PM on a Tuesday.

You press "Enter" on a complex formula. But instead of a result, your cursor transforms into a spinning blue circle. The screen washes over with that terrifying translucent white glaze. The title bar delivers the verdict: *Microsoft Excel (Not Responding).*

For the next two minutes, you don't breathe. You aren't analyzing data; you are bargaining with the universe, praying that the autosave from ten minutes ago actually worked.

This is the **Excel Ceiling**.

It is the invisible wall where the tool that built your career suddenly becomes the primary liability to your growth. To understand why you need to migrate to Power BI, we have to stop romanticizing the spreadsheet and conduct an honest audit of its structural failures.

The Hard Limit vs. The "Sluggish Zone"

Technically, since 2007, an Excel worksheet has had a binary hard limit: **1,048,576 rows**.

Back in 2007, a million rows felt like infinity. Today? In the era of high-frequency transactional logs and IoT sensors, a million rows is a shallow puddle. A mid-sized e-commerce company generates that much data in a single quarter. Once you hit row 1,048,577, Excel doesn't warn you; it simply truncates the data. It ignores reality.

But let's be real. The *theoretical* limit is irrelevant because you will never reach it. The *practical* limit is much, much lower.

The Truth No One Tells You: Excel breaks long before you hit the bottom of the sheet. Performance degradation—what we call the **"Sluggish Zone"**—kicks in anywhere between 50,000 and 300,000 rows, depending on your formula density.

Why does this happen? Because Excel is a **dependency chain engine**. If you utilize volatile functions like `VLOOKUP`, `OFFSET`, or `INDIRECT`, Excel is forced to recalculate the entire chain every time a single cell changes.

It creates a brutal loop:

- You change a date filter.
- Your laptop fan spins up like a jet engine taking off.
- You wait 45 seconds.

- You realize you selected the wrong month.
- You change it again. Another 45 seconds of your life, gone.

To survive, most analysts turn calculation to "Manual." This is a trap. It turns a dynamic analytical tool into a static snapshot. You are now flying blind, hoping you remember to hit F9 before sending that report to the CFO.

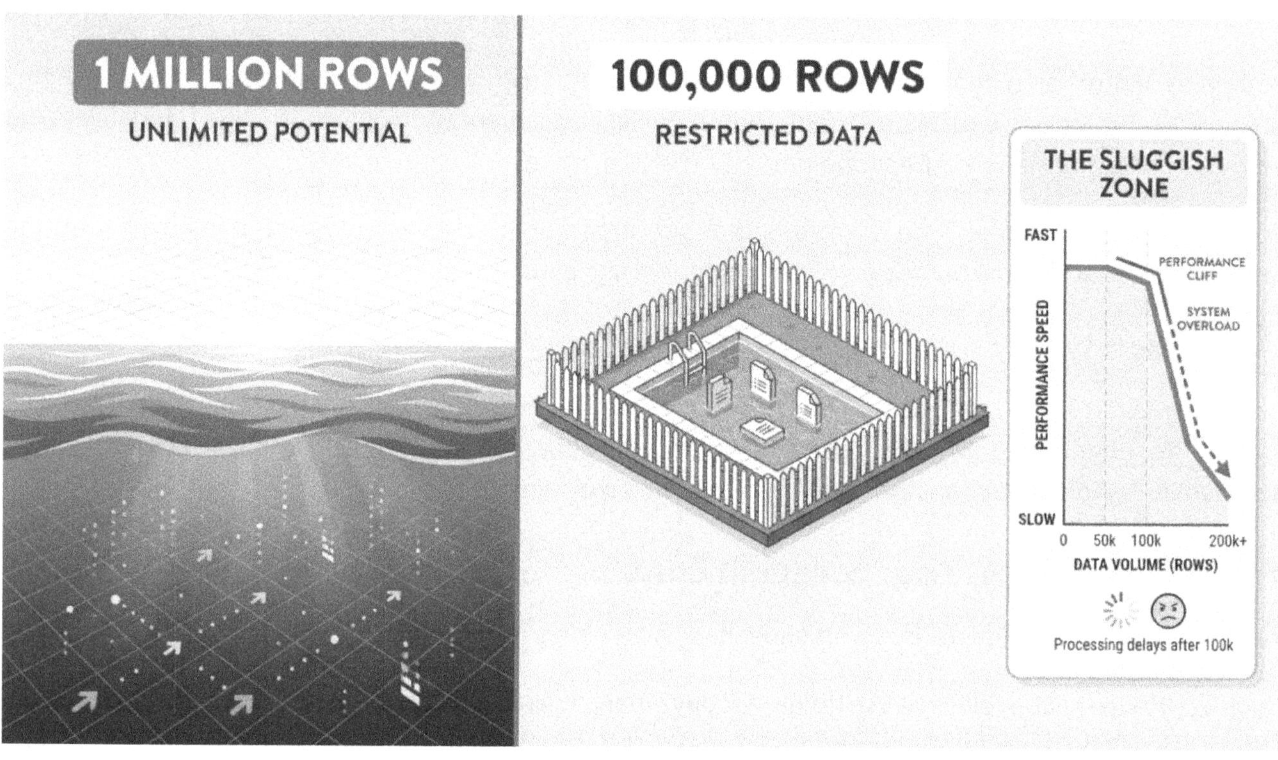

A.1 - The Excel Capacity Illusion. While the software technically supports over a million rows (The Ocean), practical performance often degrades rapidly after just 100,000 rows (The Pool). The sidebar graph illustrates the "Sluggish Zone"—the tipping point where calculation speed collapses and the tool becomes a liability.

The $6.2 Billion Copy-Paste Error

We operate under a dangerous fallacy: if the spreadsheet is slow, it's annoying, but if the numbers balance, it's safe.

It isn't. Spreadsheets lack the basic rigors of software engineering. There is no automated testing, no code review, and no systematic validation. Research by the European Spreadsheet Risk Interest Group (EuSpRIG) suggests that **88% to 90%** of all corporate spreadsheets contain errors.

Even if you are meticulous, you likely make a mistake in roughly 1% of your formula cells. When you are managing millions of dollars, 1% is catastrophic.

Take the case of the **"London Whale."**

In 2012, JPMorgan Chase lost approximately **$6.2 billion**. Analysts and pundits assumed this was due to complex derivatives trading or market unpredictability.

The Reality: It was a manual Excel error.

In the Value-at-Risk (VaR) model, a user copied and pasted a formula incorrectly. They divided by a sum instead of an average. That's it. A simple keystroke error masked the portfolio's true volatility, encouraging the bank to take on massive risks they couldn't see.

You might be thinking, *"I'm careful. I check my work."* So did the quantitative analysts at JPMorgan. These were some of the smartest financial minds on the planet. The problem wasn't their intelligence; it was the **fragility of the tool**. If you rely on manual cell referencing for mission-critical data, you aren't performing analysis. You are performing high-wire acrobatics without a net.

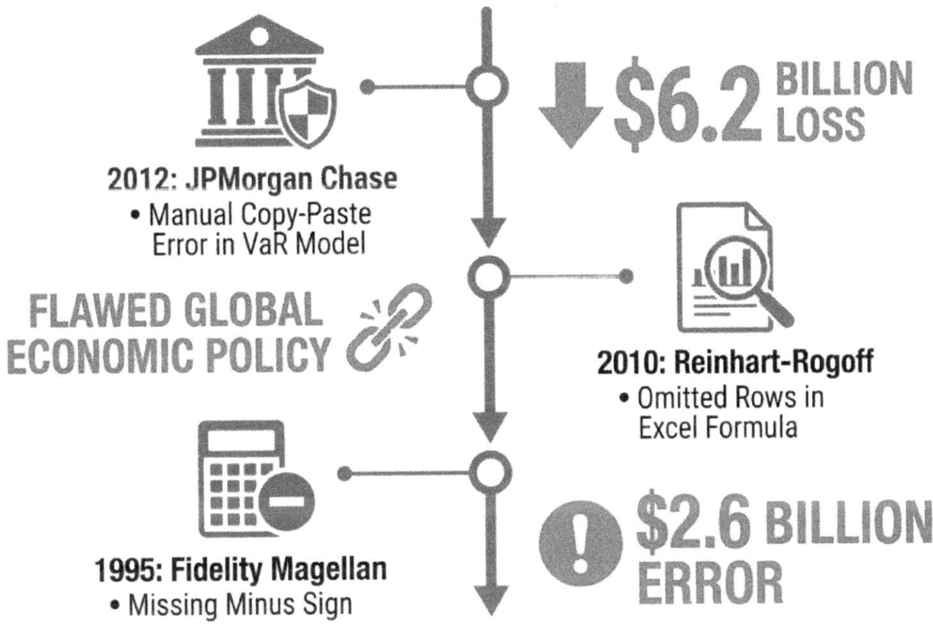

*A.2 - **The High Cost of Manual Errors**. A timeline of catastrophic financial losses, including the $6.2 billion JPMorgan incident, traced back to simple spreadsheet mistakes. These disasters highlight the inherent fragility of relying on manual cell references for mission-critical analysis.*

Operational Inefficiency: The "Version Control" Hell

Excel was designed as a personal productivity tool, not an enterprise database. It is file-based. This creates a synchronization nightmare the moment a second person needs to contribute.

You know this pain intimately. It looks like this in your folder structure.

- `Budget_2024.xlsx`
- `Budget_2024_v2.xlsx`
- `Budget_2024_v2_FINAL.xlsx`
- `Budget_2024_v2_FINAL_JK_edits.xlsx`
- `Budget_2024_v2_FINAL_REAL_DO_NOT_TOUCH.xlsx`

This is **"Version Control Hell."** It forces highly paid professionals into **"Swivel-Chair Integration"**—manually moving data from an ERP export to Excel, to another Excel file, and finally to PowerPoint.

ProcessMaker research indicates the average employee performs over **1,000 copy-paste actions per week**.

Here is the brutal truth for the aspiring leader: Every hour you spend copy-pasting data or debugging a broken `VLOOKUP` is an hour you represent a **cost** to your company, not an **asset**. You are acting as a human data-router, a job that software should be doing for free.

The Transition Point: Are You a Data Janitor?

How do you know if you have officially outgrown Excel? It isn't just about row counts. You have hit the "Excel Ceiling" if you recognize yourself in these scenarios:

1. **The History Purge:** You find yourself deleting last year's data just to make this month's data fit into the file size limit.

2. **The Frankenstein Formula:** Your `INDEX(MATCH)` logic spans across five different tabs and requires you to hold your breath when you drag it down.

3. **Key Person Risk:** You are the only person in the building who knows how to update "The File." If you get sick, the company reporting stops. (Note: This does not make you irreplaceable; it makes you **unpromotable**).

4. **The Ratio Imbalance:** You spend 80% of your time *refreshing* the report (cleaning, pasting, fixing errors) and only 20% of your time actually *analyzing* what the numbers mean.

Excel is the world's best scratchpad, but it is a terrible database. Power BI is not merely "Excel on steroids." It is a fundamental shift from *creating* data files to *modeling* data flows. It allows you to handle millions of rows with zero copy-pasting, enforcing an integrity that Excel physically cannot provide.

It is time to stop being a data janitor and start being an architect.

Mindset Shift: Moving from Cells to Columns and Contexts

The Addressability Crisis: Losing "Cell A1"

If Excel is your mother tongue, Power BI isn't just a weird dialect. It's a complete rewrite of the laws of physics.

The first panic attack happens fast. You load your data, reach for your mouse, and instinctively try to reference `C4` to tweak a formula. Then it hits you. **Cell C4 does not exist.** In Excel, the world is a grid. Every piece of data has a GPS coordinate (Column C, Row 4). It's comforting. It's tangible. In Power BI, however, data is **addressable only by name and relationship**. The grid is gone; you are now floating in a relational void.

The Logic

Power BI views data as unordered sets. This is database theory crashing into your spreadsheet reality. The engine doesn't know—and frankly, doesn't care—which row is "third" or "tenth." It only recognizes that a row belongs to the `'Sales'` table and holds a value for `[Quantity]`.

When you write DAX (Data Analysis Expressions), you are forced to abandon **procedural** thinking ("Go to A1, add it to A2"). You must adopt **declarative** thinking ("Sum the entire Quantity column, I don't care how many rows exist").

The Experience

Think of Excel as a massive warehouse with numbered aisles. You tell your assistant: "Go to Aisle 4, Shelf B, and bring me the box." Precise? Yes. But you have to know exactly where the box is.

Now, imagine Power BI as a gigantic ball pit.

You cannot tell the engine, "Give me the third ball from the left." That ball moved five seconds ago. Instead, you shout, "Give me all the *red* balls." The engine instantly grabs every red ball, weighs them, and hands you the total. You lose the ability to point at a specific spot, but you gain the ability to manipulate ten million items with a single sentence.

The Brutal Truth

Let's not sugarcoat it: **You will miss the grid.**

For the first month, it feels like working with your hands tied behind your back. In Excel, if a number looks weird, you click the cell and overwrite it. We've all done it. In Power BI, **you cannot touch the data**. There is no manual override. If the number is wrong, you must fix the source or the logic.

You are no longer a data *editor*; you have graduated to data *auditor*. This forces a level of integrity that is painful at first, but absolutely necessary for scale.

The Engine Room: VertiPaq vs. The Grid

Why suffer through this? Why trade the flexibility of a cell for the rigidity of a column?

Simple. **Speed.**

Excel runs on a grid engine built for flexibility, Power BI runs on **VertiPaq** (xVelocity), an In-memory columnar beast designed to crush massive datasets.

The Logic

Excel uses "Row-Store" logic. To sum up sales for a region, Excel often has to load the entire row (Customer, Date, Product, Region) into memory just to pluck out the "Sales" value. It reads the book line by line.

Power BI uses "Column-Store" logic. The engine slices your table vertically. To calculate "Total Sales," VertiPaq ignores the Customer, Date, and Region columns completely. It scans *only* the Sales column. Even better, it uses aggressive compression like **Dictionary Encoding**.

If you have a column with 10 million rows but only 4 unique values (North, South, East, West), VertiPaq doesn't store "North" 2.5 million times.

It stores a tiny dictionary of 4 words and a map of pointers. That's why a 500MB CSV file often shrinks to a 50MB Power BI model.

The Experience

Imagine your data is a library.

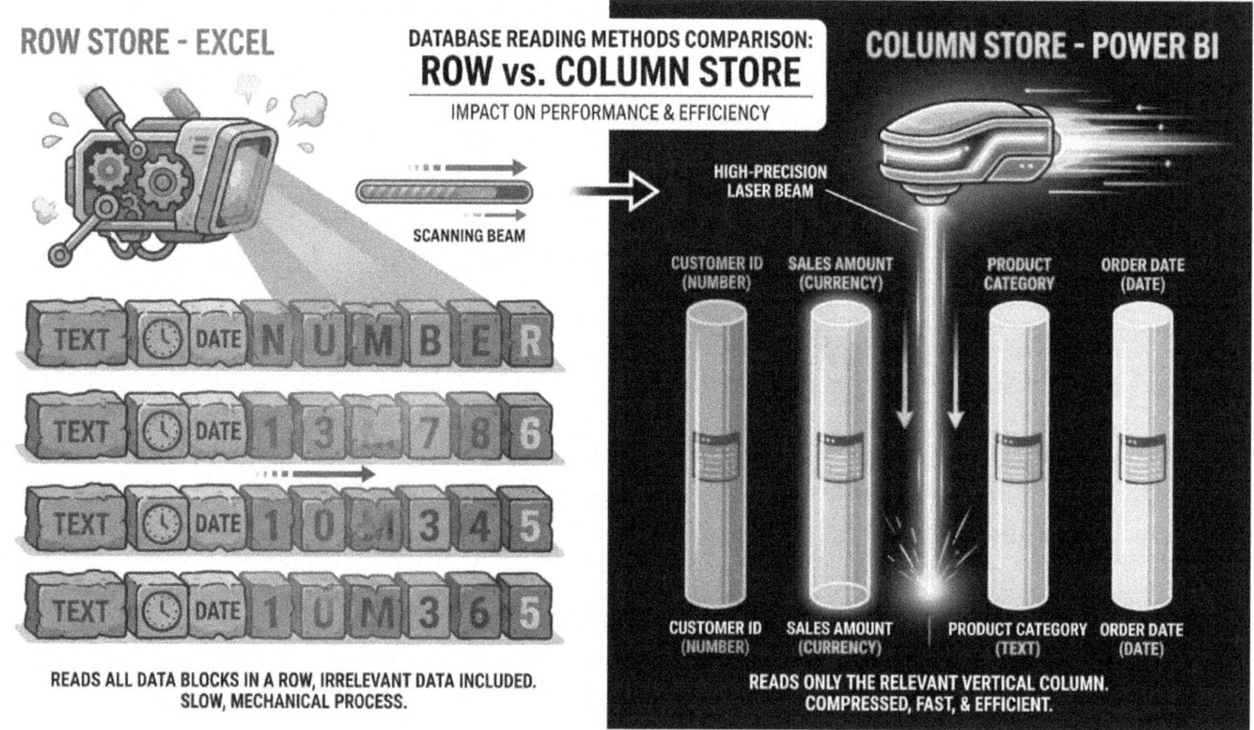

*A.3 - **The "Laser" Advantage**. Unlike Excel, which reads data line-by-line like a human reading a book, Power BI's columnar engine ignores the noise. It scans vertically, extracting only the relevant metric without touching the surrounding data blocks.*

- **Excel (Row Store):** To count the history books, you walk down the aisle, pull every single book off the shelf, check the genre, and put it back. It's exhausting.
- **Power BI (Column Store):** You don't touch the books. You walk to the index card catalog. The engine scans a simplified list that *only* contains genres. It ignores titles, authors, and page counts. It scans the "Genre" list in milliseconds.

The Brutal Truth

Power BI is fast, but it isn't magic.

Because it compresses based on *repetition*, it hates uniqueness. If you load a column with 10 million unique transaction IDs or a timestamp precise to the millisecond, compression breaks. Your model bloats.

You have to unlearn the Excel hoarder mentality of "keeping everything just in case." If a column doesn't drive a decision, it's dead weight. **Kill it.**

The Context Hurdle: Row vs. Filter

Here is the "Blue Screen of Death" moment for most Excel pros. In Excel, context is visual: the context of cell C3 is that it sits next to B3. In DAX, context is invisible, dynamic, and floats in the ether.

The Logic

You must master two invisible forces:

1. **Row Context:** This acts like Excel. It iterates row by row (e.g., "Price * Quantity"). But here's the catch—this only happens automatically in calculated columns. In measures, it doesn't exist unless you force it with specific functions like `SUMX`.

2. **Filter Context:** This is the game-changer. It refers to the active filters at the "exact moment" a calculation runs. If a user clicks a slicer for "2024," the underlying table virtually shrinks to contain only 2024 data "before" the math happens.

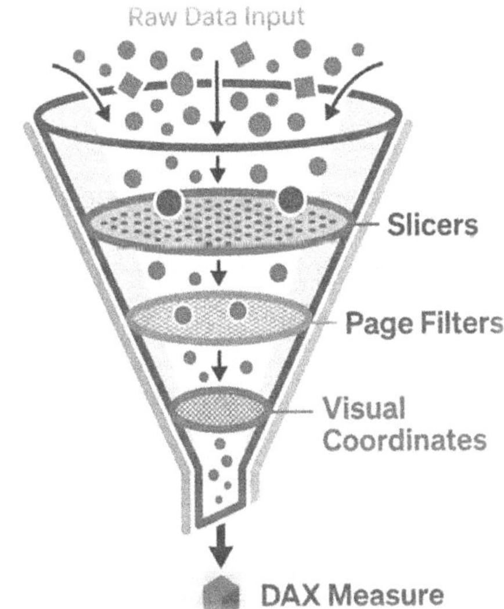

*A.4 - **The Evaluation Context Funnel.** A DAX measure is not a static calculation; it is the final result of a filtering process. Raw data is progressively whittled down by Slicers, Page Filters, and Visual Coordinates (rows/columns) before the math even begins.*

The Experience

Think of a funnel.

- **Top of the Funnel:** Your raw data (millions of rows).
- **The Filters (Slicers/Visuals):** The user clicks "USA" and "Q1." The funnel constricts.
- **The Output:** The DAX engine only sees the drops of water that make it through the funnel.

In Excel, `SUM(A:A)` always sums column A. Period. In Power BI, `SUM(Sales[Amount])` actually means: "Sum the Sales Amount... *given that the user just clicked 'USA' and 'Q1'.*"

The formula hasn't changed; the reality in which the formula exists has changed.

The Brutal Truth

You will write a measure that looks perfect. It gives the right total card. Then, you drag it into a monthly chart, and suddenly it repeats the grand total for every single month. You will scream at your monitor.

This happens because you are thinking in **cells** (static) rather than **contexts** (dynamic). Until you accept that every visual is essentially a query filtering your data in real-time, DAX will feel unpredictable. You aren't writing math; you are writing rules for how data interacts with filters.

Decoupling: The "Golden Dataset"

Finally, we have to talk about architecture. Excel files are monolithic. The data, the logic, and the pretty charts all live in one fragile `.xlsx` file. If you delete the rows, the chart dies. If you email the file to a colleague, you immediately have two versions of the truth. Chaos ensues.

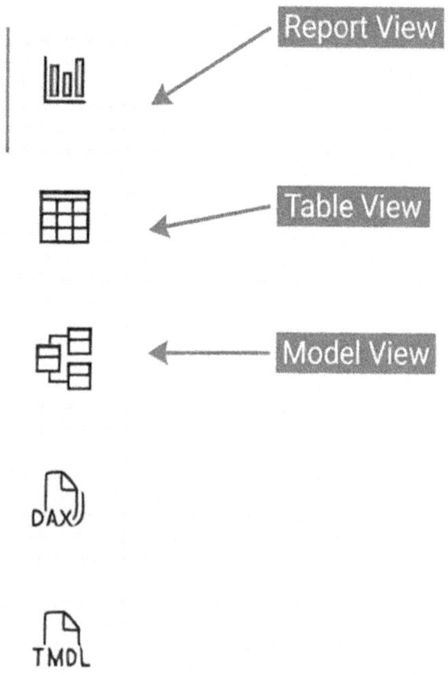

*A.5 - **The Trinity of Power BI**. Unlike Excel, where data, logic, and visuals collide in one grid, Power BI separates your workflow into three distinct environments: Report View (The Canvas), Table View (The Audit), and Model View (The Architecture).*

The Logic

Power BI separates these concerns into two distinct layers:

1. **The Semantic Model:** The invisible brain (Tables, Relationships, Measures).

2. The **Report:** The pretty face (Charts, Tables, Slicers).

This separation allows for the **"Golden Dataset"** approach. You publish one robust, certified dataset to the Cloud. That single brain can then feed 10, 20, or 50 different "thin" reports.

The Experience

In your Excel life, you are a **Data Janitor**. Every month ends with you opening ten different spreadsheets, updating the same VLOOKUPs in ten different places, and praying you didn't miss a range. In your Power BI life, you become an **Architect**. You build the foundation (the model) once. When you fix the definition of "Gross Margin" in the model, it instantly propagates to every single report connected to it. You stop moving data and start managing logic.

The Brutal Truth

This requires a massive culture shift. You can no longer "whip up" a quick report by hacking together some cells. You have to build a model first. It feels slower in the beginning. Your boss might ask, "Why is this taking two days? You used to do it in Excel in an hour." The difference? Next month, the Excel task takes another hour. The Power BI task takes **zero seconds** because the data flows automatically. You are investing time now to buy freedom later. Welcome to scalable analytics.

The Power BI Ecosystem: Desktop, Service, and Mobile

The Ecosystem: A Supply Chain for Data

To truly implement the "Golden Dataset" strategy, you must fundamentally alter your mental model of software. Stop thinking of your tools as single, self-contained executable files like Excel. **Excel is a tool; Power BI is a supply chain.** It functions as a tripartite ecosystem designed to ruthlessly separate three distinct behaviors: construction, distribution, and consumption. Understanding exactly where one ends and the other begins is the difference between a chaotic, error-prone workflow and an automated insight factory.

1. Power BI Desktop: The Architect's Studio

This is your "kitchen." It is the environment where raw ingredients—messy CSVs, complex SQL queries, and disjointed Excel sheets—are chopped, cooked, and plated.

The Why (Analyst)

Why a separate desktop application? Why not build everything in the browser like Google Sheets? **Processing power and isolation.** When you are modeling millions of rows or writing complex DAX algorithms, you need the raw compute power of your local CPU and RAM, not the latency of a web browser.

But there is a more critical reason: **Integrity.** Separating the "creation" environment from the "consumption" environment safeguards your model. In Excel, a user can accidentally type over a formula and break the entire logic chain.

In Power BI Desktop, you are the architect; the end-user is merely a visitor who cannot knock down the walls.

The Experience

Opening Power BI Desktop feels distinct from opening Excel. There is no grid of infinite empty cells waiting to be filled. Instead, you are greeted by a blank canvas and a set of controls. It is quieter here. This is deep work territory. This is where you connect to a folder of invoices, write the logic that defines "Gross Margin," and align pixels on a chart until they are perfect. The output is a `.pbix` file.

Think of this file not as a spreadsheet, but as a compressed, encrypted engine. It contains the data cache, the semantic model, and the visual layer all in one artifact.

The Reality Check

Let's be honest: Power BI Desktop is powerful, but it is ugly. It looks like Microsoft Office 2013 had a bad day. The ribbon is cluttered, and the formula bar for DAX is nowhere near as user-friendly as a modern code editor (though it's improving).

Mac users, brace yourselves: You are currently out of luck. There is no native Mac version of Power BI Desktop. You will need to run a Virtual Machine (like Parallels) or use a cloud PC. It adds friction, but for the power it offers, it is the tax you must pay.

Also, unlike Excel, there is no "Undo" button for data refresh steps—so build carefully.

2. Power BI Service: The Publishing House

Once the meal is cooked, you do not invite customers into the kitchen to eat out of the pans. You send the food to the dining room. Power BI Service (`app.powerbi.com`) is that dining room.

The Why (Analyst)

The Service is a SaaS (Software as a Service) platform designed to eliminate the "Version Control Nightmare." In the old world, you emailed `Report_v2_FINAL.xlsx` and prayed everyone opened the right one.

In the Power BI ecosystem, you publish the `.pbix` file to the Service.

This creates a single **Golden Dataset** in the cloud. Crucially, this allows for automation.

You can schedule refreshes. If your data lives on a local server, you use a piece of software called a **Gateway**—a bridge that tunnels through your corporate firewall to push local data up to the cloud securely.

The Experience

Imagine waking up, grabbing your coffee, and opening your browser to find that your report has already updated itself. The data from yesterday's close is there. No copy-pasting, no refreshing pivots, no anxiety about whether you captured the right range.

This is where the concept of the **App** comes in. You can bundle your reports into a slick, navigational "App" that functions like an internal website for your executive team. They don't see the tables or the messy backend; they just see the insights, clean and ready for decision-making.

The Terminology Trap

You need to be careful here, because Microsoft makes it confusing.

- In Excel, you call your summary sheet a "Dashboard."
- In Power BI, a **Report** is the interactive, multi-page deep dive you built in Desktop.
- A **Dashboard** is a specific technical feature in the Service—a single scrolling page where you "pin" charts from different reports.

The brutal truth? Most professionals rarely use the "Dashboard" feature. It is often too high-level and lacks the interactivity (slicers/filters) that analysts require. Stick to sharing **Reports** or **Apps** unless your CEO specifically demands a single-pane-of-glass monitor for the lobby TV.

Also, be aware of the "Freemium" trap. Desktop is free. Publishing to the Service for others to see usually requires a **Pro license** (for both you and the viewer). Budget accordingly.

3. Power BI Mobile: The Executive Brief

If the Desktop is for the Analyst and the Service is for the Manager, the Mobile App is for the Executive who is running between meetings.

The Why (Analyst)

Information loses value over time. If a Regional Manager has to wait until they return to the office to check sales figures, the opportunity to correct a behavior might be lost. The Mobile App extends the "Golden Dataset" to the edge of the network.

The Experience

This is the "elevator pitch" device. You are in a hallway, and the VP of Sales asks, *"How are we tracking against the Q3 target?"* In the Excel era, you would say, *"I'll check and email you."* In the Power BI era, you pull out your phone. The report is already there, formatted specifically for the vertical screen. You tap a bar on the chart, highlight an anomaly with your finger, and text the screenshot directly to the team. You look competent, prepared, and in control.

The Warning

Do not assume your desktop report will "just work" on mobile. By default, Power BI takes your landscape desktop view and turns it sideways. It looks terrible and is impossible to read. To make this work, you must spend extra time in the **Mobile Layout** view in Power BI Desktop, rearranging your visuals for a phone screen. If you skip this step—and most lazy developers do—your users will delete the app within a week because "it's too hard to read."

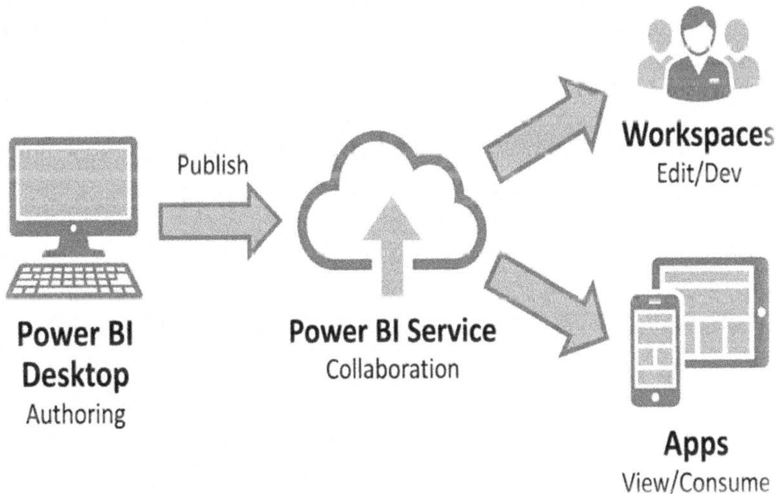

*A.6 - **The Publishing Lifecycle.** The journey from local Authoring (Desktop) to Cloud Collaboration (Service). Note the critical distinction at the end: Workspaces are for the development team (Editors), while Apps provide a secure, read-only experience for business users (Viewers).*

The Cycle of Efficiency

The transition from Excel to the Power BI ecosystem is a shift in your personal labor economics.

1. **Ingest & Model (Desktop):** You invest heavy hours here upfront.
2. **Publish & Schedule (Service):** You hand off the maintenance to the cloud.
3. **Consume & Decide (Mobile/Web):** You and your stakeholders reap the dividends.

Your goal as a modern analyst is to spend **80% of your time in the Desktop phase** building a robust model, so that the Service phase runs on autopilot. You are moving from being a "Data Janitor" who cleans the same spreadsheet every Monday, to a "Data Architect" who builds the machine that does the cleaning for you.

Installation and Setup: Preparing Your Analytics Workbench

The "Workbench" Concept

If Excel is a sketchpad, Power BI Desktop is a heavy-duty industrial workbench.

That isn't a metaphor. It is a technical reality. In Excel, you open a file, type in a cell, and save. It is light. It is agile. It is forgiving. Power BI is fundamentally different. When you launch the application, you aren't just opening a document editor; you are spinning up a local instance of Microsoft's SQL Server Analysis Services engine right on your laptop.

Think of this transition like moving from a handheld calculator to a CAD workstation. You are no longer just "entering data." You are engineering a data product.

This environment requires a shift in mindset—from distinct, disconnected spreadsheets to a centralized, structural development environment. You are becoming an architect of information. And an architect requires a stable foundation.

Hardware Reality Check: Official vs. Real World

Here is the truth that Microsoft's marketing page won't tell you. The official minimum specifications are designed for *viewing* simple reports, not for *building* the complex analytical models you are here to learn. When you load data into Power BI, it utilizes an engine called **VertiPaq**. This engine compresses data and stores it entirely in your computer's RAM (Random Access Memory) for lightning-fast retrieval.

If you rely on the official "minimum" specs, your experience will consist mostly of staring at a spinning loading circle while your machine gasps for air.

In Excel, performance is often bound by calculation chains. In Power BI, performance is bound by memory availability. Every column you import consumes RAM. If you show up to a data fight with 4GB of RAM, you have already lost. You need headroom for the Operating System, your browser (which likely eats RAM for breakfast), and the Power BI engine itself.

Component	Microsoft Official Minimum	The Real-World Requirement
RAM	4 GB	**16 GB+**
CPU	1 GHz	**2 GHz+ Multi-core**
Storage	1 GB	**SSD / NVMe**

*A.7 - **Official Specs vs. Reality**. Microsoft's minimum requirements are designed for viewing simple reports, not building complex models. Because Power BI's VertiPaq engine runs entirely in-memory, prioritizing RAM (16GB+) and fast storage (SSD) is the single best investment you can make to prevent the "spinning wheel of death."*

The Operating System Barrier

This is often the hardest pill for creative professionals and marketing analysts to swallow.

Power BI is a Windows-exclusive ecosystem.

As of 2026, there is no native Mac version. There likely never will be due to the deep dependencies on the Windows .NET framework and COM components. If you are a Mac user, you are facing a massive friction point. You have three choices, ranked by viability:

1. **Parallels Desktop (The Pragmatic Choice):** This software allows you to run Windows virtually inside macOS. It creates a "Coherence" mode where Power BI looks like it's running on your Mac dock. It is seamless, but it is resource-intensive. Your Mac needs to be powerful enough to run two operating systems simultaneously without choking.

2. **Cloud VM (The Laggy Choice):** You can rent a Windows PC in the cloud via Azure or AWS. However, developing over a remote desktop connection introduces a micro-second lag between your mouse movement and the cursor response. For a "high efficiency/low friction" worker, this latency is torture.

3. **The Hard Truth:** If you are serious about pivoting your career into data analytics, the path of least resistance is to buy a Windows laptop. It sounds harsh. But fighting the OS compatibility battle every day drains the energy you should be using to solve data problems.

Installation Method: Store vs. Exe

There are two ways to install Power BI Desktop: downloading the `.exe` file from the web, or installing it via the **Microsoft Store**. While IT departments often prefer the `.exe` for control, you should fight for the Store version.

Why? **The "Update Treadmill."**

Microsoft updates Power BI Desktop *every single month*. They add new DAX functions, new visualization types, and critical performance patches.

- **The Store Version:** Updates automatically in the background. You wake up, and the new features are there.
- **The .exe Version:** Requires you to manually download and reinstall the software every month.

Then there is the **Compatibility Trap**. Power BI files are not backward compatible. If your colleague opens a file with the June version and saves it, you cannot open it with your May version.

You will get a "Version Incompatible" error, and your workflow will grind to a halt. The Store version ensures you are always in sync with the ecosystem.

The Licensing "Gotcha"

Let's clarify the economics immediately, because they are confusing.

- **Power BI Desktop is Free:** You do not need a license to install the software, import 50 million rows, write complex code, and build stunning dashboards. It is free to create.
- **Power BI Service is Paid:** The moment you want to share that dashboard securely with your manager via the cloud, the meter starts running. You (and the viewer) generally need a **Pro** license (roughly $10/user/month).

Think of it this way: You can write the book for free (Desktop), but if you want to publish it to the library where others can read it (Service), you have to pay the distributor. For learning purposes, the free Desktop version is all you need right now.

Critical First Configuration: Kill "Auto Date/Time"

Before you load a single byte of data, you must change one default setting. This is the hallmark of a professional setup versus an amateur one.

The feature is called **"Auto Date/Time."**

By default, Power BI scans your data for any column that looks like a date. It then secretly generates a hidden date table for every single one of those columns. If you have a sales table with "Order Date," "Ship Date," and "Due Date," Power BI builds three hidden tables in the background.

The cost? This bloats your file size—often increasing it by 30% to 50%—and slows down performance. It is a "training wheels" feature designed to help beginners who don't know how to model time.

You are here to take the training wheels off.

The Fix:

1. Go to **File > Options and settings > Options**.

2. Under **Current File**, select **Data Load**.

3. **Uncheck** "Auto Date/Time".

This forces you to build a single, centralized Date Table (which we will cover in Chapter 4), ensuring your model is lean, fast, and accurate.

The "Work Email" Hurdle

Finally, you may hit a wall when trying to sign in. Microsoft requires a "work or school" account (e.g., `@company.com`) to log in to the Power BI Service. If you try to use `@gmail.com` or `@outlook.com`, the door will slam shut.

If you are learning independently and don't want to use your current employer's domain, do not give up. There is an insider workaround.

Join the **Microsoft 365 Developer Program**.

Microsoft will give you a free, renewable "sandbox" tenant. This comes with a legitimate organizational domain (e.g., `@yourname.onmicrosoft.com`) and 25 E5 licenses. You get a fully functional, enterprise-grade environment to practice publishing and sharing reports, completely bypassing the personal email restriction. This is how you build a portfolio without a corporate sponsor.

The Roadmap: A Preview of the Case Study Project

The Mission: Welcome to SmartGear Retail

Enough with the abstract theory. You cannot learn to swim by reading a textbook on fluid dynamics, and you certainly cannot master Power BI by memorizing the location of buttons on a ribbon. To truly understand this tool, you need a job to do. For the rest of this journey, you are stepping into the role of **Lead Analyst** at "**SmartGear Retail.**"

Picture a mid-sized electronics distributor that is currently drowning in its own success. Revenue is up, but visibility is zero. The CEO knows how much money the company made last year, but if she asks a specific question—"Which product category drove the highest margin in Q3 specifically in the Northeast region?"—the current team freezes. They stare at their shoes. They mumble, "Give us two days to pull the numbers."

Your mission is to retire that two-day delay.

You are going to dismantle their fragile, manual reporting infrastructure and replace it with a fully automated **Executive Sales Dashboard**.

The Raw Material: Anatomy of "Excel Hell"

Your starting point is a digital crime scene. It will feel uncomfortably familiar. In fact, it is the exact environment found in about 90% of corporate finance and operations departments today.

Here is what we are dealing with:

- **The Fragmented History:** Sales transactions aren't in one place. They are severed limbs scattered across folders: `Sales_2023.csv`, `Sales_2024.csv`, and—dreadfully—`Sales_2025_Jan_v2.csv`.
- **The Static Silo:** Product costs and categories are locked inside a local file named `Products_Master.xlsx` on someone's desktop.
- **The CRM Export:** Customer details live on a third disconnected island, a `Customer_List.csv` export that hasn't been updated in weeks.

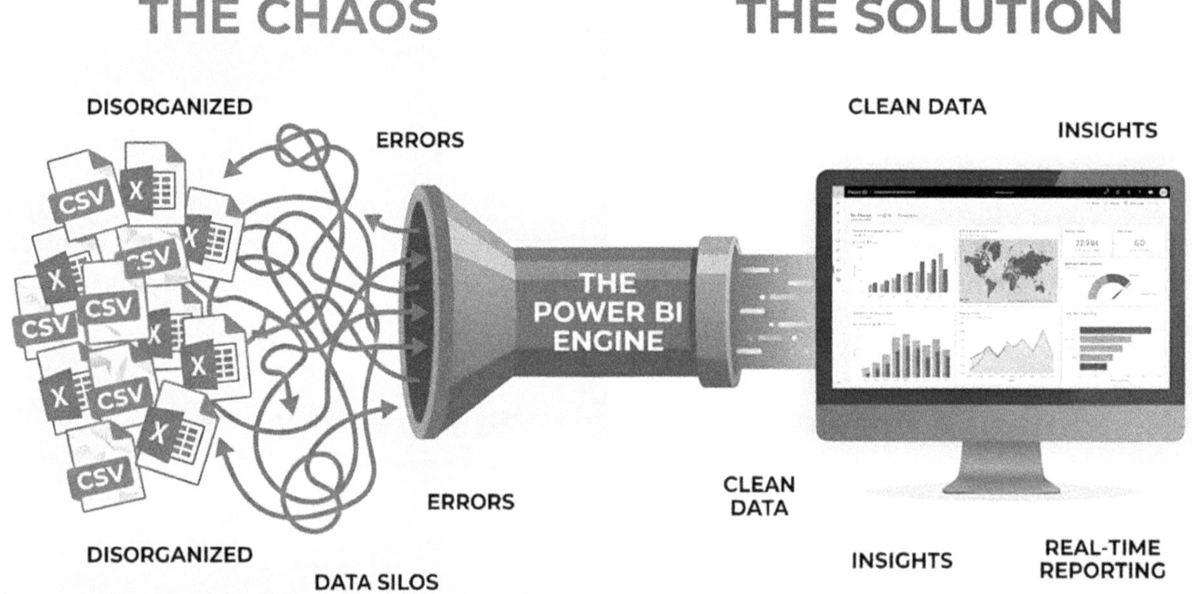

*A.8 - **From Data Janitor to Architect.** The shift from manual "Data Janitor" work—fighting tangled CSVs and broken links (Left)—to automated "Data Architecture" (Right). The Power BI Engine doesn't just display data; it cleans and stabilizes it, removing human error from the loop.*

Let's be honest about why this setup fails. It isn't just "messy"; it is a massive corporate liability. Every time you open that Master Excel file to paste in new rows, you roll the dice. You risk a copy-paste error. You risk overwriting a formula. You risk breaking a VLOOKUP range. Worst of all, you are wasting your intelligence on **data janitor work**—cleaning rows—instead of actual analysis.

We are going to stop that. Immediately.

The 5-Phase Implementation Plan

We will execute this turnaround using the industry-standard **Business Intelligence Lifecycle**. This isn't just a checklist; it is the architectural difference between a spreadsheet that breaks and an application that scales.

Phase 1: Data Connection (The Extraction)

The Goal: Establish a "live wire" to the data.

The Shift: In Excel, you *possess* the data; you physically paste it into the grid. In Power BI, you *connect* to it. We will teach you to point Power BI not at a specific file, but at a **Folder**.

The Payoff: This is the secret to automation. When January 2026 rolls around, you won't touch the report. You will simply drop `Sales_2026.csv` into that folder. Power BI will detect the new file, ingest it, and update every chart automatically. No copy-paste. No human touch.

Phase 2: Data Transformation (The Kitchen)

The Tool: Power Query Editor.

The Goal: Preparation. Think of raw data like raw produce—it's dirty and needs chopping.

The Action: We will standardize column names, fix date formats, and filter out errors.

The Truth: This is the unsexy part that nobody talks about, yet it consumes 70% of a typical analyst's time. Power BI records your cleaning steps like a macro on steroids. You do the cleanup once, and the system replays those steps every time the data refreshes. We are automating the grunt work.

Phase 3: Data Modeling (The Brain)

The Tool: The Model View.

The Goal: Constructing a **Star Schema**.

The Shift: This is where Excel users usually crash. Your instinct is to mash everything into one giant "flat" table using VLOOKUP to bring it all together. **Do not do this.** It creates a bloated, slow model.

The Action: We will keep our tables separate (Sales in one table, Products in another) and connect them via **Relationships**. This architecture is what allows you to slice 10 million rows of data in milliseconds without crashing your CPU.

Phase 4: Intelligence (The Calculations)

The Tool: DAX (Data Analysis Expressions).

The Goal: Defining the metrics that matter.

The Action: We will write **Measures—**portable formulas like `Total Revenue`, `Profit Margin %`, and `YoY Growth`.

The Concept: Unlike an Excel formula which lives in Cell C2 and refers to Cell B2, a Measure is a virtual concept. It floats above the data. If you drag `Profit Margin %` onto a map, it calculates by geography. Drag it onto a timeline, it calculates by month. It is context-aware logic.

Phase 5: Visualization & Distribution (The Story)

The Tool: The Canvas & Power BI Service.

The Goal: Self-Service BI.

The Action: We will build the dashboard using interactive Slicers, Maps, and Matrixes, and then publish it to the Cloud.

The End of the Attachment: You will no longer email a 15MB file on Monday mornings. You will send a link. When the CEO opens that link on her iPad, she is looking at the "Single Source of Truth." She can filter by Region or Product herself.

The Promise

By the end of this guide, you won't just have a chart. You will have a machine. A machine that ingests raw chaos and spits out decision-ready insights, allowing you to stop answering the question, **"What happened?"** so you can finally start answering, **"What should we do next?"**

Day 1 - Data Preparation: Killing the Copy-Paste Routine

1.1 Connecting to Data: Excel, CSV, and Web Sources

The Paradigm Shift: From Ownership to Access

In the Excel world, your relationship with a file is possessive. You open it, you lock it, you own it. If you double-click `Sales_2023.xlsx`, you are effectively holding that data hostage. It consumes your RAM. It creates a bottleneck where no one else can edit the file. It is a static, manual existence.

Power BI changes the physics of this relationship.

You do not "open" data anymore; you **connect** to it.

Think of Excel as buying a physical book and scribbling notes in the margins. You own that specific copy, and if you lose it, the information is gone. Power BI, conversely, acts like a library index. It reads the book, notes the location of the information, and creates a reference system. It never writes in the book itself.

This distinction is not just philosophical; it is vital for three practical reasons:

1. **Integrity:** You cannot accidentally delete a row in your source system through Power BI. The source remains pristine.

2. **Scalability:** Power BI uses the VertiPaq engine. It reads the data and compresses it into a cache that is often **10x smaller** than the original Excel file.

3. **Live Connection:** When your colleague adds March sales to the source file, Power BI doesn't need a new file. You simply hit "Refresh," and it re-reads the library.

Your command center is the **Home > Get Data** menu. While the dropdown shows the usual suspects like Excel and SQL Server, the "More..." option reveals a library of over 100 connectors, bridging everything from simple text files to complex cloud databases like Salesforce.

Connector 1: Excel Workbook

To start, select **Get Data > Excel Workbook** and navigate to your local file. It seems simple enough. But immediately after selecting the file, you hit the most critical decision point in the entire import process: **The Navigator Window.**

You will likely see your data listed twice. Once with a "Sheet" icon and once with a "Table" icon.

Here is the truth that separates pros from amateurs: Connecting to a *Sheet* is asking for suffering.

Excel Sheets are unstructured canvases. They are messy. They contain "creative" elements: a logo in cell A1, a title in A2, a blank row in A3, and perhaps a "Note: These numbers are draft" typed in column Z. If you connect to the Sheet, Power BI imports *all* of this noise. You will waste the next hour writing logic to remove top rows and filter out nulls.

The Solution: Always connect to **Excel Tables** (Ctrl+T in Excel).

A Table is a structured object. It tells Power BI exactly where the data starts and ends. It automatically promotes headers. It ignores the notes in the margins. If you control the source file, force it into a Table format before you ever open Power BI. It is the single highest-return efficiency hack for Excel data sources.

Connector 2: Text/CSV

If an Excel file is a luxury sedan with heated seats (formatting, formulas, colors), a CSV file is a dragster. It is stripped of all comfort for the sake of raw speed. When you are dealing with datasets exceeding **500,000 rows**, abandon Excel files.

The XML structure inside an `.xlsx` file creates significant overhead during the read process. A CSV (Comma Separated Values) file creates zero formatting overhead. In benchmarking, Power BI can ingest CSV data up to **6x faster** than equivalent Excel data.

However, because CSVs are "dumb" text files, you must be the brains during the import:

1. **File Origin:** Power BI usually guesses "1252 (Western European)" or "65001 (UTF-8)". Unless you see strange symbols replacing accents, leave this alone.

2. **Delimiter:** It will auto-detect commas or semicolons. Give this a one-second glance to ensure your columns are split correctly.

3. **Data Type Detection (The Trap):** By default, Power BI guesses data types based on the **first 200 rows**.

The Trap: Imagine a "Termination Date" column. For active employees, the first 200 rows are empty. Power BI looks at them, sees nothing, and sets the column type to "Empty" or "Text". Later, in row 5,000, a date appears. Power BI will trigger an error because that date violates the "Empty" rule it established.

The Fix: If you suspect a column is sparse (mostly empty at the top), do not rely on auto-detection. Fix the data types manually in the next stage (Power Query).

Connector 3: Web (Public & SharePoint)

The Web connector allows you to treat the internet—and your corporate intranet—as a database.

Scenario A: Public Web Scraping

If you need exchange rates or a list of US states, don't copy-paste from Wikipedia. Use **Get Data > Web** and paste the URL. Power BI scans the HTML specifically looking for `<table>` tags. In the Navigator, switch to the **Web View** tab. It offers a "what you see is what you get" experience; click the table on the webpage, and Power BI grabs the data.

Scenario B: SharePoint/OneDrive (The "Direct Path" Hack)

This is where 90% of users get stuck.

You have a file on SharePoint. You try the "SharePoint Folder" connector, but it tries to sync thousands of files from the entire site. You just want *one* file.

Here is the secret handshake to connect to a single cloud-hosted Excel file efficiently:

1. Open the file in your **Excel Desktop App** (not the browser).
2. Go to **File > Info**.
3. Click **Copy Path**.
4. Go to Power BI, but select the **Web** connector (Yes, "Web," not "SharePoint").
5. Paste the link.
6. **CRITICAL:** The link will end with `?web=1`. You **must delete** these last six characters.

Why? The `?web=1` suffix forces the link to open in a web browser renderer. By deleting it, you point Power BI directly to the raw file binary. If you skip this step, the connection will fail.

Authentication Levels: The Gatekeeper

When you connect to a new source, Power BI acts like a bouncer. It needs to know who you are before letting you in. You will see a prompt with several tabs on the left. Choosing the wrong tab is the most common cause of "Access Forbidden" errors.

- **Anonymous:** Use this *only* for public web pages.
- **Windows:** Use this for files on your local hard drive or mapped network drives (e.g., the `Z:` drive). It uses your current laptop login.
- **Organizational Account:** This is the **only** correct choice for SharePoint, OneDrive, or Microsoft 365 data. Even if you are logged into Windows, SharePoint requires a specific OAuth2 token passed via this option. You must click "Sign In" here.

Transition to Transformation: The "Load" Button is Lava

Once you have selected your tables, the Navigator window offers two buttons at the bottom: **"Load"** and **"Transform Data"**.

Your instinct, driven by years of wanting to "get the job done," will be to click "Load".

Do not do it.

Clicking "Load" imports the data exactly as it is—dirty, unstructured, and heavy. It immediately compresses errors into your model, making them harder to find later.

Always click "Transform Data".

Think of this like cooking. Clicking "Load" is like throwing unwashed vegetables, raw meat, and eggshells directly onto the dinner plates. Clicking "Transform Data" takes you to the kitchen—the **Power Query Editor**—where we wash, peel, and chop the ingredients before serving them.

We are now entering the kitchen. Welcome to Power Query.

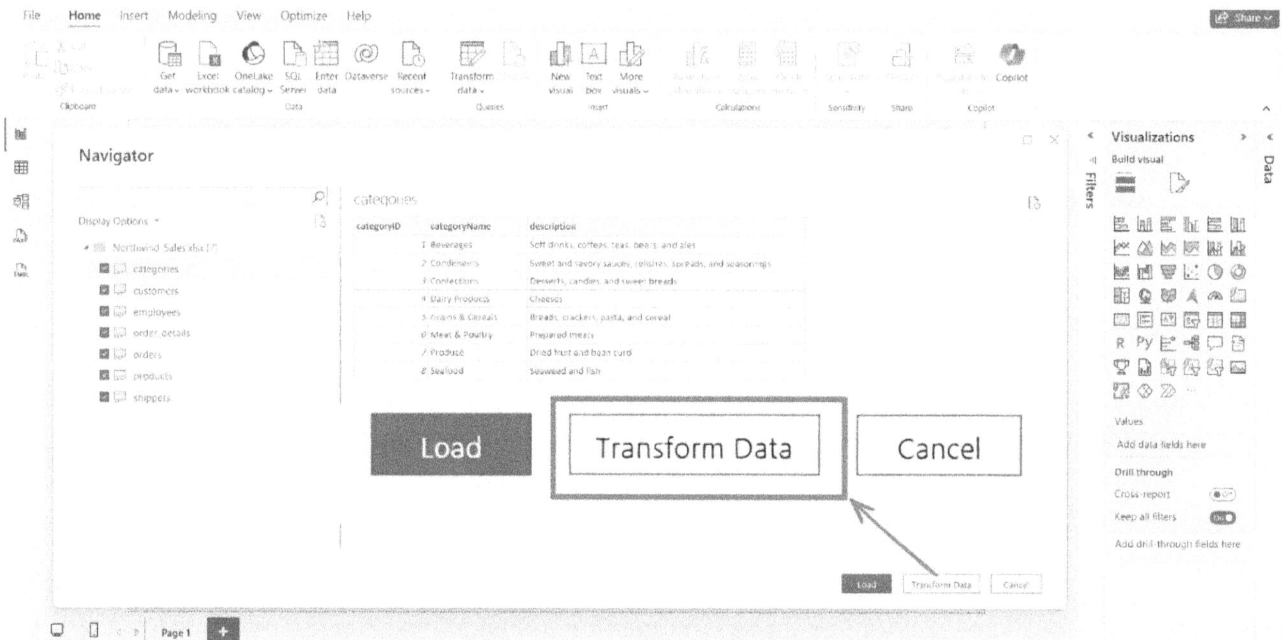

*1.1 - **Entering the Kitchen.** Clicking Load dumps raw, messy data directly into your report. Clicking **Transform Data** (highlighted) opens the Power Query Editor, allowing you to wash, peel, and chop your data ingredients before serving them to the dashboard.*

1.2 The Query Editor Interface: Your New Automation Engine

The Environment: A Separate Reality

When you click "Transform Data," watch what happens. The screen shifts. You aren't just opening a sub-menu; you are crossing a threshold.

Power Query Editor is a modal window—a distinct application layering over Power BI. Think of the main Power BI interface as the **Showroom**: polished, interactive, client-ready. Power Query is the **Engine Room**. It's louder. It's industrial. It's where the actual torque is generated. While you are in this window, you are in a "sandbox." You cannot interact with your charts.

You can't tweak colors. Good. This isolation is intentional. It forces a necessary discipline: structure first, aesthetics later.

The Analyst's Insight:

Why the wall between the two? Speed. The data grid you see here isn't your live SQL database. It is a **cached preview**. Usually just the top 1,000 rows. Delete a column here, and your source file remains untouched. You are simply writing a recipe, not carving the meal. Power BI downloads a small sample to let you design your process without waiting for millions of rows to load every time you click a button.

DAY 1

The Four Quadrants of Power Query

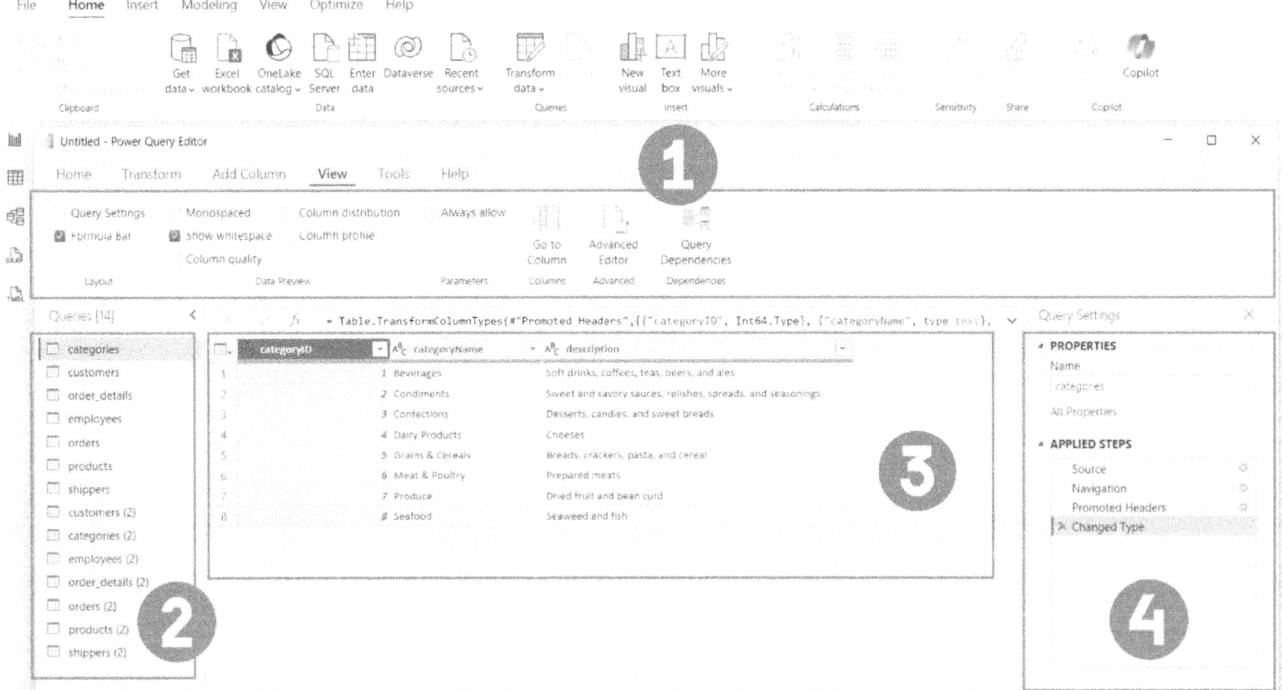

1.2 - ***Navigating the Editor.*** *(1) The Ribbon contains your transformation tools for shaping data; (2) The Queries Pane acts as a navigation bar for your connected tables; (3) The Data Preview displays the current state of your data (remember: you cannot edit cells directly here); and (4) Query Settings tracks your "Applied Steps," serving as both your recipe and your undo history.*

To command this engine, you need to understand the four distinct control panels.

1. The Ribbon: The Command Center

At a glance, this looks like Excel. Don't be fooled. In Excel, you change data cells. Here, you change *logic*.

The Efficiency Trap:

There is a specific mistake that catches 90% of beginners. Look closely at the tabs. You will see two options with nearly identical buttons: **"Transform"** and **"Add Column."**

- **Transform:** This modifies the column *in place*. Turn "lowercase" text to "UPPERCASE," and the original data is destroyed. Replaced.

- **Add Column:** This creates a *brand new* column with the uppercase text, leaving the original column sitting next to it.
- **The Truth:** Always default to "Transform." "Add Column" doubles the memory required for that specific field. Only use "Add Column" if you logically need to compare the old state against the new. Otherwise, you are just bloating your model.

2. The Queries Pane: Your Inventory

On the left, you see every table you've connected to.

The Architect's Advice:

Rename your queries. Now. By default, Power BI is lazy. It pulls in names like `Sheet1` or `Table_004`. If you leave these, you will be lost by Tuesday. Press **F2** and give them descriptive, boring names like `Fact_Sales` or `Dim_Customers`. Your future self—tired, stressed, and debugging a report in six months—will thank you.

3. The Data Preview: The Map, Not the Territory

The center screen shows your grid of data. It feels like Excel, but it isn't editable. You can't double-click a cell and type "Steve." You must apply a *rule* to change values.

The Reality Check:

Look at the bottom left corner. It likely says "Column profiling based on top 1000 rows."

This distinction is crucial: **You are looking at a sample.** If your dataset has ten million rows, and the only error hides in row 9,000,050, you will *not* see it here. Do not assume clean data just because the preview looks perfect. This view is for designing logic, not auditing every single record.

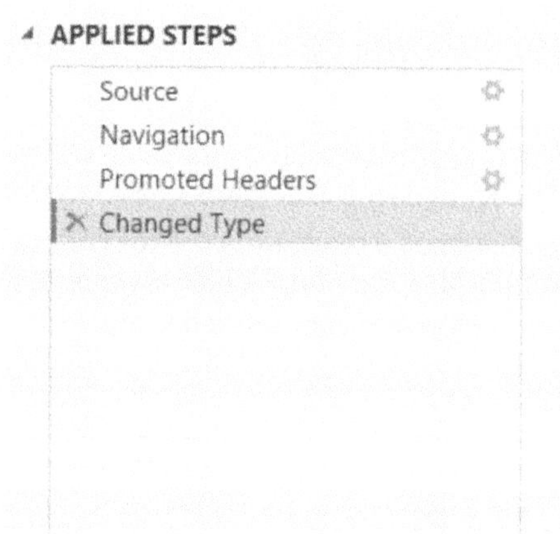

*1.3 - **The Automation Script.** Think of the **Applied Steps** pane as a video recorder for your data cleaning. It captures every click in sequence—from Source to Removed Columns. When you refresh next month, Power Query simply hits "Play" and repeats the work for you.*

4. The Applied Steps: The Automation Engine

On the right side lies the "Query Settings" pane. This is the heart of the beast.

The Engine: Applied Steps

In Excel, you act in the "now." You delete a row, and it's gone. You run a VLOOKUP, and the result sits there. If you make a mistake, you mash Ctrl+Z and pray you didn't save the file.Power Query is different. It is a **Time Machine**.

Every time you click a button—remove a column, filter a date, replace a value—Power Query doesn't just do it; it *records* it. These are the **Applied Steps**.

The Storyteller's Perspective:

Imagine you are filming a tutorial on how to clean a spreadsheet. You hit record, do the work, and stop. Next month, when the new data file arrives, you don't do the work again. You just hit "Play" on the video. Power Query is that video player. It applies your steps, in exact order, every single time you refresh the data.

How to Navigate Time:

- **Chronological Dependency:** Steps execute from top to bottom. If Step 3 creates a "Total Cost" column, and Step 4 filters that column, you cannot delete Step 3 without crashing Step 4.
- **The Gear Icon:** See a gear icon next to a step? That means the logic is editable. Click it to change your mind (e.g., change a filter from "Sales > 100" to "Sales > 500") without rewriting the whole script.
- **The "X":** The red X next to a step is your delete button. But be careful—this isn't just an "Undo"; it's a structural removal of a line of code.

The Hidden Layer: "M" Code

If you don't see a formula bar above your data, go to **View > Formula Bar** immediately. Turn it on.

*1.4 - **View > Formula Bar:** Turn this on immediately. It transforms your button clicks into visible logic, showing you exactly how Power BI translates your actions into the M scripting language.*

When you click buttons in the ribbon, you are actually using a User Interface (UI) writer that generates a script in a language called **M** (Mashup).

The Analyst's "Why":

You don't need to learn to write M code today. However, you need to look at it. Why? Because it demystifies the magic. When you filter a row, you aren't performing manual labor; you are generating a function: `= Table.SelectRows(...)`. Recognizing this shift—from manual worker to script architect—is the moment you stop being an Excel user and start being a Data Developer.

Quality Control: Data Profiling

Excel users are used to "silent failures"—errors that hide until a VLOOKUP returns ` #N/A` or a Pivot Table sum looks weirdly low. Power Query has sensors to catch these issues before they enter your report.

Action: Go to the **View** tab. Check **"Column Quality"** and **"Column Distribution"**.

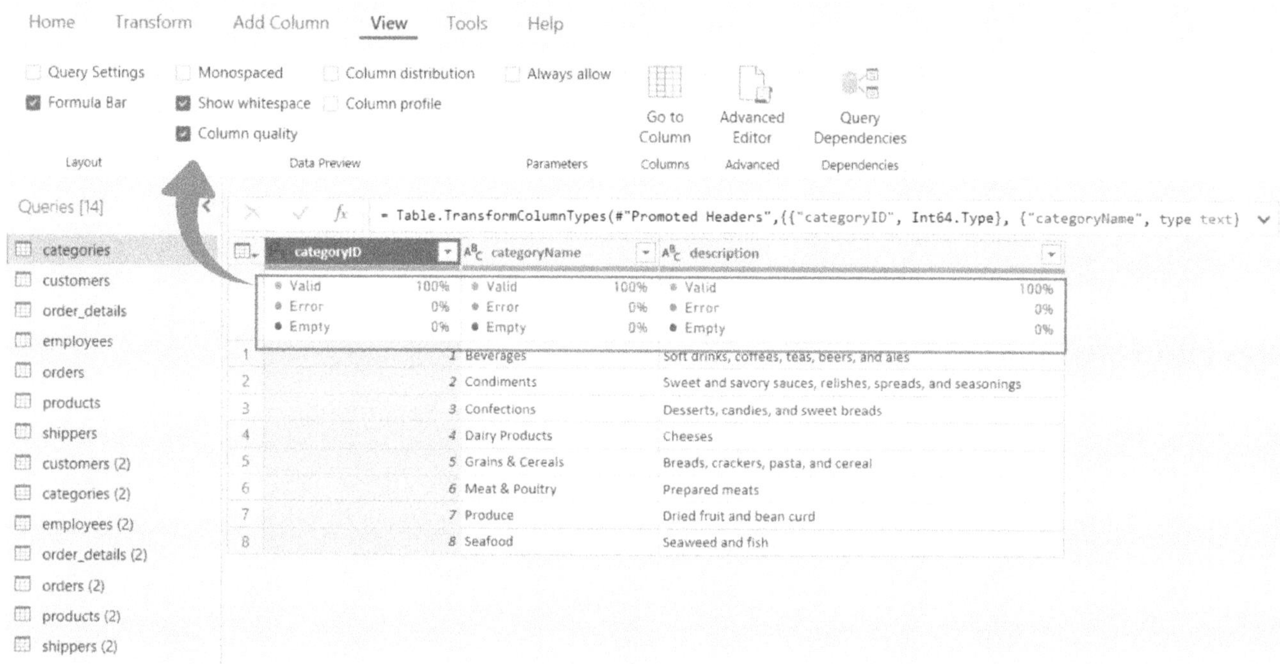

*1.5- **The Data X-Ray.** Never scroll through a million rows to find errors. Just look at the "Health Bar" under the column header. **Green** is good, **Red** is a syntax error (e.g., text in a number column), and **Grey** is empty data (nulls).*

You will see a colored bar appear under every column header. This is your dashboard health monitor:

- **Valid (Green):** The data matches the column type (e.g., numbers in a number column).
- **Error (Red):** The data contradicts the rule (e.g., the text "N/A" inside a Date column). In Excel, this cell would just look weird. In Power BI, this will cause your data refresh to crash.
- **Empty (Gray):** These are null values.

The Final Word:

Never overlook the color indicators. If you spot even a tiny sliver of red—even just 1%—stop immediately. Do not pass Go. Take a moment to hover your cursor over the red bar to investigate exactly what went wrong.

Addressing these errors right here at the source will save you countless hours of frustrating troubleshooting and debugging broken reports later on.

In the high-stakes world of data automation, a healthy dose of paranoia isn't just a trait; it's a virtue.

1.3 Essential Transformations: Cleaning, Splitting, and Formatting

Excel operates like a digital sketchpad—it tolerates ambiguity. You can write a date as text, add stray spaces for alignment, or put three different concepts in a single cell. Power BI, however, operates like a strict database. It demands absolute precision.

DAY 1

If Excel is the "Wild West," Power Query is the sheriff that imposes law and order.

The transition from Excel to Power BI is often painful because of one specific realization: **Human data entry is messy.** Your VLOOKUPs didn't break because the formula was wrong; they broke because "Product A" (clean) does not match " Product A " (messy).

In this section, we move from simply connecting to data to performing structural surgery on it.

1. The Invisible Enemies: Trim, Clean, and the "Ghost" Space

In the realm of databases, empty space is not empty data. It is a character.

To a computer, the string `"Product A "` (with a trailing space) is mathematically distinct from `"Product A"`. If you try to link these two tables in your Data Model, the relationship will fail silently. You won't get an error message; you will just get blank results in your visuals.

The Standard Hygiene Routine:

- **Trim (`Text.Trim`):** This is your first line of defense. It acts like a laser, slicing off *leading* and *trailing* whitespace. Crucially, it leaves the single spaces *between* words alone.
- **Clean (`Text.Clean`):** This removes non-printable control characters—the invisible debris of the digital world. These are Line Feeds, Tabs, and Carriage Returns that often tag along when you copy-paste data from PDFs or legacy web systems (like SAP or Salesforce). You can't always see them, but they will break your logic. **The "Ghost" Character (The Truth No One Tells You)**

Here is a scenario that drives analysts insane. You look at a cell. It looks empty. You apply `Trim`. It still looks empty. But Power BI insists there is data there.

You have likely encountered the **Non-Breaking Space** (ASCII 160 or Unicode `00A0`).

This is common in web-scraped data or exports from HTML-based systems. It is *visually* a space, but *programmatically* a different character entirely. Standard `Text.Trim` only targets the standard space (ASCII 32). It will not touch the Non-Breaking Space.

The Fix: If `Trim` fails you, do not panic.

1. Select the column.
2. Choose **Replace Values**.

3. In the "Value To Find" box, you cannot type this character easily. You may need to copy it from the original bad cell or use a special code.

4. In "Replace With", hit the spacebar once (standard space).

5. *Then* run your Trim operation.

2. Structural Surgery: Splitting Columns (Rows vs. Columns)

Excel users are addicted to "Text to Columns." It feels natural to split a list like `"High Priority, Pending, Q1"` into three columns: *Tag 1*, *Tag 2*, and *Tag 3*.

In Power BI, this creates a nightmare.

If you split into columns, how do you answer the question: *"How many projects are High Priority?"* You would have to write a complex DAX formula to sum up counts from *Tag 1*, *Tag 2*, and *Tag 3*. This is bad modeling. It creates "Wide Data" which is slow and hard to manage.

The "Split to Rows" Revolution

Power Query offers a hidden gem that transforms your data from "Wide" to "Tall" (normalized).

- **The Scenario:** You have a Sales ID column and a Tags column: `ID: 101 | Tags: "Red, Large"`
- **The Action:** Select **Split Column > By Delimiter**.
- **The Secret Step:** Expand **Advanced Options** and select **"Rows"** (instead of Columns).

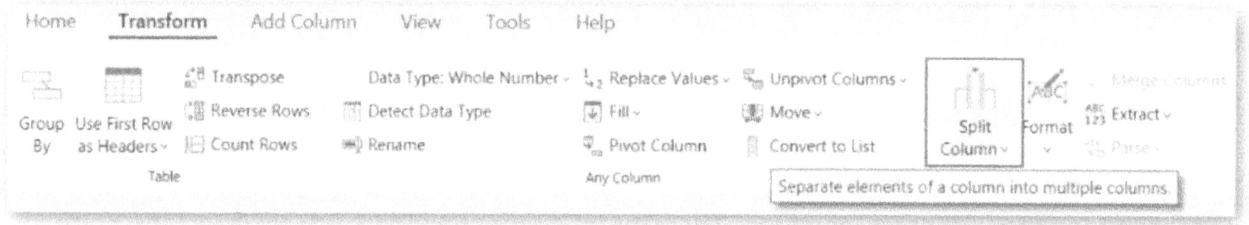

*1.6 - **The Vertical Transformation.** By expanding Advanced options and selecting Rows, you force Power Query to normalize the data. Instead of creating messy new columns, it generates a clean vertical list—perfect for Slicers and aggregation.*

The Result: Power Query duplicates the Sales ID for each tag.

- Row 1: `ID: 101 | Tag: Red`
- Row 2: `ID: 101 | Tag: Large`

Now, you can simply drag the "Tag" field into a Slicer or a Bar Chart. Power BI can effortlessly count, filter, and aggregate by tag because the data is structurally vertical. You have traded a messy horizontal list for a clean vertical table—the gold standard of data modeling.

3. The Rosetta Stone: Formatting & Locale

New users often treat Data Types (Text, Date, Decimal) as cosmetic formatting, similar to changing font colors in Excel.

The Truth: Defining a Data Type is an assertion of fact.

When you tell Power Query a column is a "Date," you are unlocking the powerful Time Intelligence engine (e.g., `SAMEPERIODLASTYEAR`). If that column remains "Text," those functions are dead on arrival.

The Regional Trap

Here is where the "Global" economy breaks your local report.

- **The Conflict:** Your laptop is set to **US English** (Month/Day/Year). You import a CSV file generated in London or Sydney (Day/Month/Year).
- **The Error:** You see `25/01/2026`. Power Query reads this on your US machine as the "25th Month." Since that doesn't exist, it returns an **Error**.

The Fix: "Using Locale"

Many analysts try to fix this by hacking the system settings or writing complex text-parsing formulas. Stop. Power Query has a built-in translator.

Do **not** just click the icon to change the type to Date. That applies *your* PC's logic to the file. Instead:

1. Right-click the column header.
2. Select **Change Type > Using Locale...**
3. **The Logic:** You are not asking Power BI to *convert* the data yet; you are telling it the **Origin Story** of the data. You select "Date" and "English (United Kingdom)."

You are effectively issuing a translation command: 'Interpret this incoming text using **British rules** (Day first, Month second), but store it internally as a universal **Date object**.' This distinction is what separates fragile spreadsheets from robust data models. By defining the *source logic* explicitly, you make your query bulletproof. It no longer matters if a colleague in New York (MM/DD) or a cloud server in Berlin (DD.MM) refreshes the dataset. The date value remains absolute and correct, while the local display adapts automatically to the user's machine settings.

1.4 The Pivot Paradox: Why Unpivoting is Crucial for Analysis

Let's be honest about how your brain works after years of staring at spreadsheets. You are wired to love "Wide Format" data.It feels natural. You want to see *January, February, March* stretching across the top like a calendar. It mimics the printed reports we've been handing to executives for decades. Your eye scans left-to-right, and everything makes sense.

But here is the uncomfortable truth that keeps analysts working late nights: **The format that is easiest for you to read is the absolute hardest for Power BI to process.**

We call this the Pivot Paradox. By structuring your source data for human consumption, you are inadvertently strangling the database engine. Power BI—and frankly, any modern analytics platform—doesn't read like a human. It operates on vector-based processing. It craves vertical streams of data, not horizontal headers. When you feed it a wide Excel table, you are forcing a square peg into a round hole. The friction this causes? It doesn't just slow down your report. It forces you to write nightmare formulas to compensate for bad architecture.

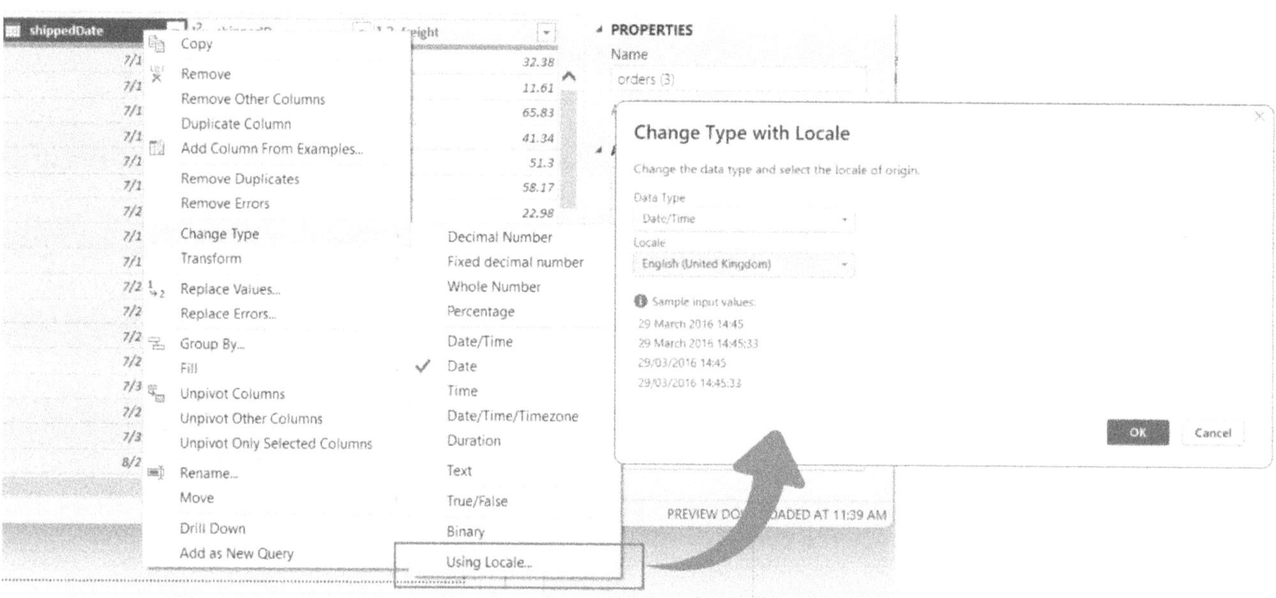

*1.7 - **The Date Translator.** If your source file uses DD/MM/YYYY (UK/Europe) but your PC expects MM/DD/YYYY (US), Power BI will either generate errors or swap the months and days. "Using Locale" forces the engine to interpret the source format correctly before converting it.*

The "Tidy Data" Mandate

To stop fighting the software and start leveraging it, we have to adopt the **"Tidy Data"** principle. Forget the academic definition for a moment. In the context of business intelligence, it means your data must follow three non-negotiable laws if you want your model to work:

1. **Each variable is a column.** (Crucial distinction: "Month" is a variable. "January" is a data point).
2. **Each observation is a row.**
3. **Each cell holds exactly one value.**

Look at your current budget file. If "January" is a column header, you have broken Law #1. In a proper dataset, "January" belongs *inside* a column named `[Month]`.

This is the fundamental difference between a report (output) and a dataset (input).

The Mathematical Cost of Being stubborn

Why does this matter? Let's look at the math. If you load that budget table with 12 separate columns for Jan–Dec, calculating a simple Year-to-Date total becomes a brittle, ugly exercise. You have to write a DAX formula that looks like this:

` [Jan] + [Feb] + [Mar] + ... + [Dec]`

It is tedious. It breaks easily. And god help you if the fiscal year changes structure.

However, if you **Unpivot** that table into a "Tall" format—creating just one column for `[Month]` and one for `[Amount]`—that monstrosity of a formula collapses into a single, elegant command:

`SUM([Amount])`

Suddenly, you aren't hard-coding column names. You are defining logic. The machine filters the `[Month]` column based on context, and the math just works.

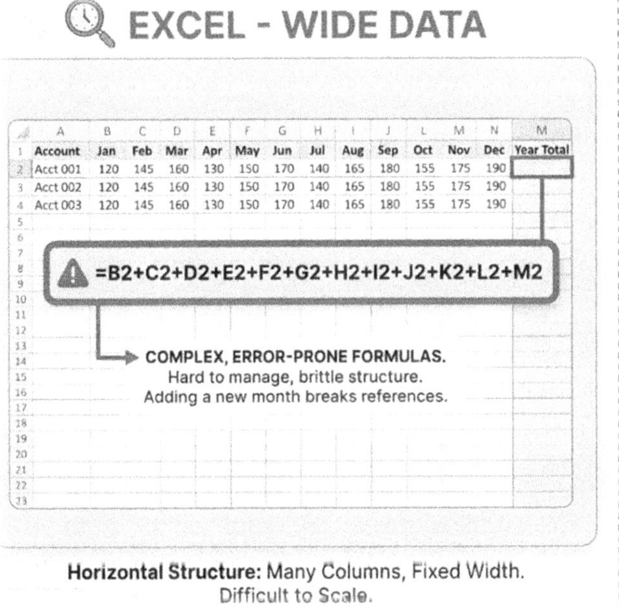

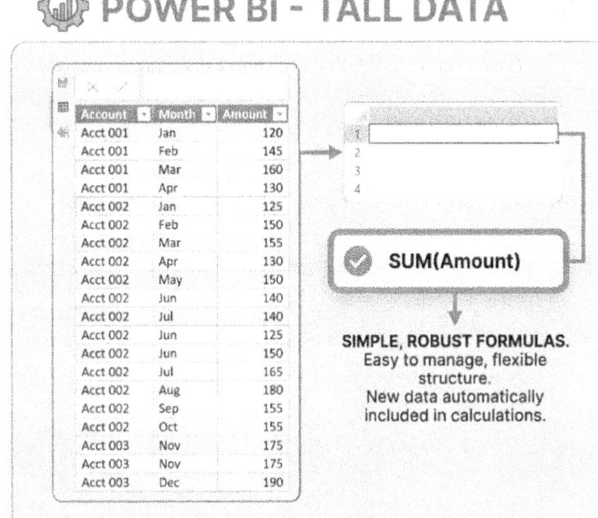

*1.8 - **The Great Shift.** Moving from "Spreadsheet Hell"—where fragile links break constantly (Left)—to a robust Star Schema (Right). Power BI doesn't just display your data; it restructures it into a professional database that updates automatically.*

The "Unpivot Other" Trap

Now, pay attention. This is where 90% of self-taught Power BI users create a time bomb in their model.

When you bring your data into Power Query to fix this, you will see the columns for *Jan, Feb, Mar*, etc. The temptation is to highlight those month columns, right-click, and hit "Unpivot Columns."

Do not do this.

If you explicitly select your month columns and unpivot them, you are creating a static rule. You are telling Power Query: *"Transform January through December."*

DAY 1

But what happens next year?

When your source file updates and "January 2026" appears, Power Query will ignore it. Why? Because you didn't tell it to look for that specific column name. Your data pipeline will silently fail, missing the new data while reporting zero errors. You won't know until a stakeholder points out the numbers look low.

The Protocol: Define the Anchors

The correct approach is counter-intuitive. Instead of telling the system what to move, tell it what to **keep**.

1. Select your **Anchor Columns** (the identifiers that must never move, like *Department*, *Account ID*, or *Region*).
2. Right-click these headers.
3. Select **Unpivot Other Columns**.

See the difference? You are creating resilient logic. You are telling the engine: *"Keep the Department and Account ID fixed, and take literally everything else—no matter how many new months get added in the future—and flip them into rows."*

This turns a fragile, static table into a dynamic data stream.

The Kitchen vs. The Dining Room

There is a psychological hurdle here you need to clear.

When you finish unpivoting, the result will look "wrong" to your Excel-trained eye. The table will be narrow, incredibly long, and repetitive. You will see the same *Account ID* listed 12 times—once for each month.

You will feel a visceral urge to pivot it back. **Resist that urge.**

Think of your data model as a commercial **kitchen**. The kitchen is not designed for the customer's comfort; it is designed for the chef's efficiency. It is messy, hot, and industrial. That is your tall data. It is optimized for the machine to crunch millions of rows in milliseconds.The report—the charts, the matrices, the dashboards—that is the **dining room**. That is where you present the data. You can use a Matrix visual to pivot the data back into a readable, horizontal format for your manager. But never confuse the two. Keep the back end tall and tidy. Let the machine do the heavy lifting so you don't have to.

1.5 Building Robust Queries: Best Practices for Future-Proofing

The Whitelist Principle: Choose, Don't Remove

You have likely spent your entire career playing defense. You open an Excel file, spot the garbage data, select it, delete it, and save. It feels productive because the mess disappears. You are the janitor of your spreadsheets. But in Power Query, that instinct is a liability. When you build a query, you aren't cleaning data for *today*; you are engineering a machine that must clean data for *next year*.

A robust query handles changes in source files without crashing. A fragile query implodes the moment your IT department changes a single column header or adds a legal disclaimer to an export. Here is how to build a machine that doesn't break.

Stop Blacklisting

When you need to get rid of columns you don't use, your muscle memory screams "Delete." You see a column named `temp_calc_v1`, you right-click, and you hit **Remove Column**. This is "Blacklisting." You are defining what you *don't* want. The problem with blacklisting is that it assumes you know every piece of garbage that will ever exist. It is a reactive, losing strategy. If the source system updates next month and replaces `temp_calc_v1` with `temp_calc_v2`, your query will blindly import this new garbage column because you never told it not to. Your model bloats, performance tanks, and you won't know why until it's too late.

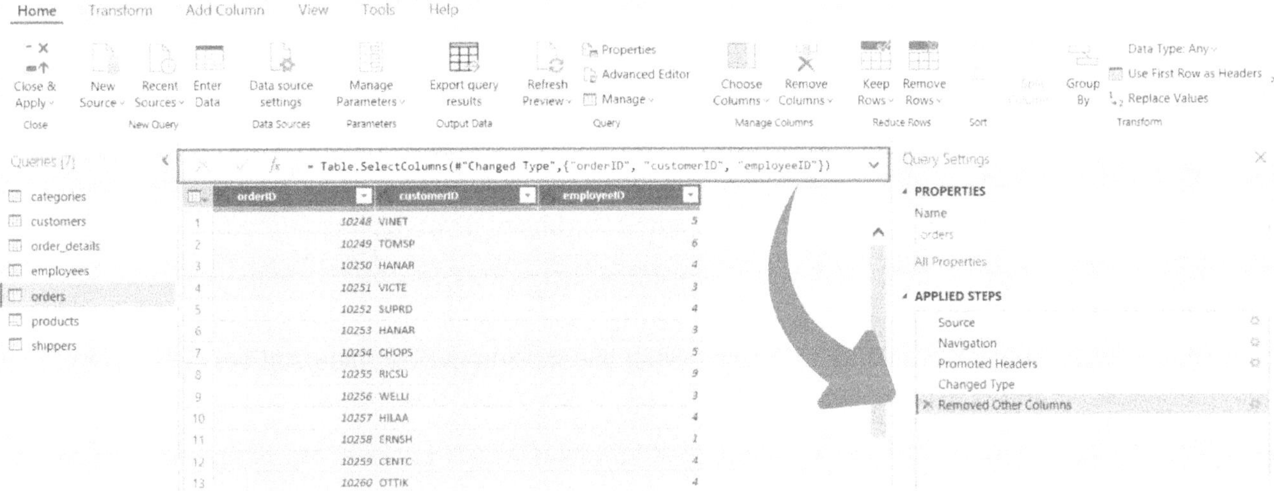

*1.9 - **The Stability Rule:** Never define what you want to remove. Always define what you want to keep. "Whitelisting" ensures that when your source file changes unexpectedly, your report doesn't break.*

Think of it like being a bouncer at an exclusive club.

- **Blacklisting (Remove Columns):** You stand at the door with a list of three specific people who are banned. Everyone else gets in. If a new troublemaker shows up who isn't on your list, they walk right in and ruin the party.
- **Whitelisting (Select Columns):** You stand at the door with a VIP guest list. You only let in `Date`, `Customer`, and `Amount`. It doesn't matter if the troublemakers change their disguises or if a hundred new ones show up. If they aren't on the list, they stay out.

The Fix:

Invert your logic. Never simply remove a column. Instead, use **Choose Columns** (from the Home ribbon) or right-click the columns you actually need and select **Remove Other Columns**. You are explicitly defining the "Survivors." This guarantees that your model remains pristine, regardless of the chaos happening upstream.

DAY 1

The "Changed Type" Minefield

Power BI tries to be helpful. Often, it tries too hard.

By default, immediately after you load a file, the engine inserts a step called **"Changed Type"**. It hardcodes the data format (text, number, date) for every single column it detects. For a novice, this is convenient. For a professional, it is a saboteur. This step creates a "hard dependency" on column names very early in your process. If your query involves dynamic transformations—like unpivoting columns where names might shift—having a hardcoded type step at the beginning acts like a concrete wall.

If a column name changes slightly before this step, the query returns a fatal error: ` Expression.Error: The column '[Name]' of the table wasn't found. Imagine you are renovating a house. The "Changed Type" step is like pouring quick-drying concrete around your plumbing pipes before you've even decided where the sink goes. If you realize later that you need to move the sink (pivot the data), you have to jackhammer the concrete.

The Fix:

Delete the automatic "Changed Type" step if it references columns that might change or be unpivoted.

1. **Keep it fluid:** Perform all your structural changes (filtering, unpivoting, merging) while the data is still "loose."

2. **Type at the Anchor:** Only apply data types **at the very end** of your query, specifically to the final columns you are loading into the model. This ensures that intermediate steps don't break just because a column header shifted.

Future-Proofing Metadata Rows

Corporate exports are rarely clean tables. They usually arrive with "header junk"—top rows containing titles like "Generated by SAP v4.2" or "Confidential - Internal Use Only." The amateur move is to use **"Remove Top Rows"** (e.g., Remove Top 3). This is a **Positional Reference**. It relies on the physical layout of the Excel sheet remaining static forever.

If the IT department updates the report and the disclaimer grows from 3 rows to 4 rows, your query will delete the first 3, and then try to promote Row 4 (the remaining junk row) as your headers. Your column names become garbage, and your refresh fails. Using "Remove Top Rows" is like walking through your living room blindfolded and counting exactly five steps to reach the couch. It works fine until someone moves the couch six inches to the left. Then, you trip.

The Fix:
Stop counting rows. Start looking for patterns. Use **Dynamic Filtering**.

Instead of saying "Delete rows 1 through 3," look at the data. Usually, the junk rows will have `null` or blank values in columns where actual data should be.

- *Fragile:* Remove Top 3 Rows.
- *Robust:* Filter the *Order_Date* column to remove "nulls" or blanks.

By filtering based on **content** (logic) rather than **position** (luck), your query becomes immune to layout changes. Even if the system adds 50 rows of legal disclaimers next month, your filter simply sees them as "not dates" and wipes them away, leaving your actual headers intact.

DAY 1

1.6 Quick Start Project: From Zero to Dashboard (Northwind Sales)

We have spent this entire chapter in the "Kitchen" (Power Query), washing and chopping our data. Now, it is finally time to enter the "Dining Room" (Report View) and plate the dish. In this final section of Day 1, we are going to build your first real dashboard. This exercise corresponds exactly to the **Day 1 Video - Module 1.4: "Quick Win: First Visual"**. If you want to follow along click-by-click, open the video now. We will follow the exact same sequence.

Phase 1: The Import (Getting the Ingredients)

1. Open a fresh **Power BI Desktop** file.
2. On the **Home** ribbon, click **Get Data > Excel Workbook**.
3. Navigate to your folder and select the `Northwind_Sales_Data.xlsx` file. Click **Open**.

The Navigator Check:

A window appears showing the tables inside the file.

1. Check the box next to the table (e.g., `Orders` or `Sales`).
2. **CRITICAL STEP:** Do **not** click "Load" yet. Click **Transform Data**.
   - *Why?* As seen in the video, raw data often contains empty rows ("Nulls") or wrong formats. If we load now, we load garbage. We must clean it first.

Phase 2: The "Sanity Check" (Cleaning Nulls & Types)

You are now in the Power Query Editor. The video (Min 03:26) emphasizes a specific cleaning routine to prevent relationship errors later.

Step 1: The "Null" Purge

If your data has empty rows (Nulls), it can break your calculations and relationships.

1. Go to the first column (e.g., `OrderID` or the primary key).
2. Click the **Drop-Down Arrow** next to the column header.
3. Look at the filter list. Scroll down and **Uncheck (null)** or **(blank)**.
4. Click **OK**.

- *Result:* You have just filtered out "dirty" incomplete rows. The video explains that removing nulls is mandatory for dimension tables to work correctly.

Step 2: The Data Type Audit

Look at the icons to the left of the headers.

- **OrderDate:** Must have a **Calendar** icon (Date). If it says "ABC", click the icon and change to **Date.**
- **Revenue:** Must have a **Decimal (1.2)** icon.
- **Product/Country:** Must have an **ABC** (Text) icon.

Step 3: Rename

In the **Query Settings** pane (right side), change the name from `Sheet1` to **`Northwind Sales`**.

Step 4: The Switch

Now the data is clean.

- Click **Close & Apply** (Top Left).
- Power BI processes the data and takes you to the blank **Report Canvas.**

Phase 3: Building the Visuals

Now we build the dashboard.

Visual 1: The Breakdown (Bar Chart)

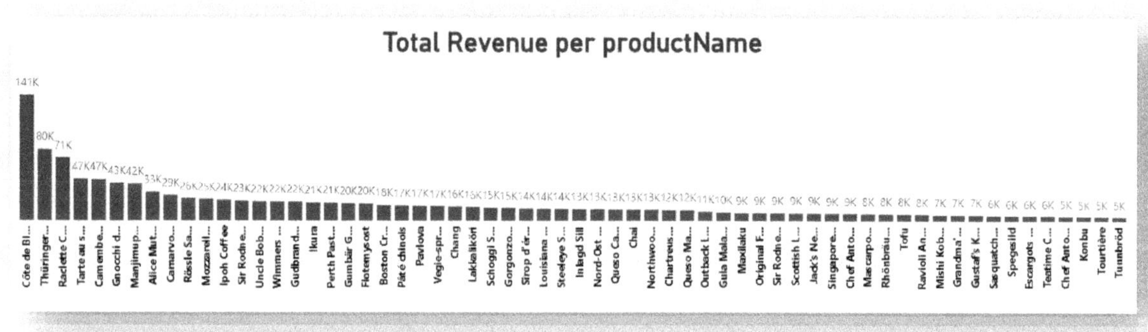

Business Question: Which products are driving revenue?

1. Open the **Visualizations Pane.**
2. Click **Clustered Column Chart.**
3. Drag **`ProductName`** to the **X-Axis.**
4. Drag **`Revenue`** to the **Y-Axis.**

- *Result:* An instant ranking of your products.

The "Blank Space" Rule:

Always click on the **white empty canvas** to deselect your chart before picking the next one.

Visual 2: The Geography (Map)

Business Question: Where are our customers?

1. Click the blank canvas.
2. Click the **Map** icon (Globe).
3. Drag **`City`** to **Location**.
4. Drag **`Revenue`** to **Bubble Size**.

Visual 3: The Trend (Line Chart)

Business Question: When are we selling?

1. Click the blank canvas.
2. Click the **Line Chart** icon.
3. Drag **`OrderDate`** to the **X-Axis**.
4. Drag **`Revenue`** to the **Y-Axis**.

The Hierarchy Drill-Down:

You likely see a single dot for the "Year". To see the months:

1. Look at the visual header (top right of the chart).
2. Click the **"Pitchfork" icon** (double arrow splitting). This is **Expand All Down**.
3. Click it twice to go from Year > Quarter > Month.

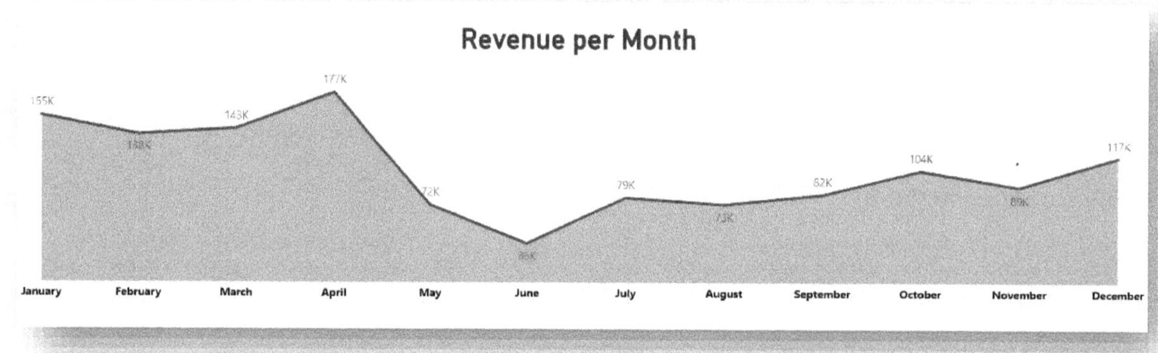

Visual 4: The KPI (Card)

1. Click the blank canvas.
2. Click the **Card** icon ("123").
3. Drag `` `Revenue` `` into the field.
    - *Result:* Your total sales number.

Visual 5: The Interaction (Slicer)

city
Tutte

1. Click the blank canvas.
2. Click the **Slicer** icon (Funnel).
3. Drag `` `City` `` into the field.
4. *Format Tip:* In the Format pane, change Style to **Dropdown**.

Phase 4: Formatting (The "Polish")

We finish by adding a professional header.

1. Go to **Insert > Shapes > Rectangle**.
2. Draw a bar at the top of the report.
3. In the Format pane, change color to Blue.
4. Turn on **Text** and write **"Northwind Sales Report"**.
5. Check your interactions by clicking a bar in the first chart. Watch the map and card update instantly.

Below is the result you should have by applying all the indications correctly.

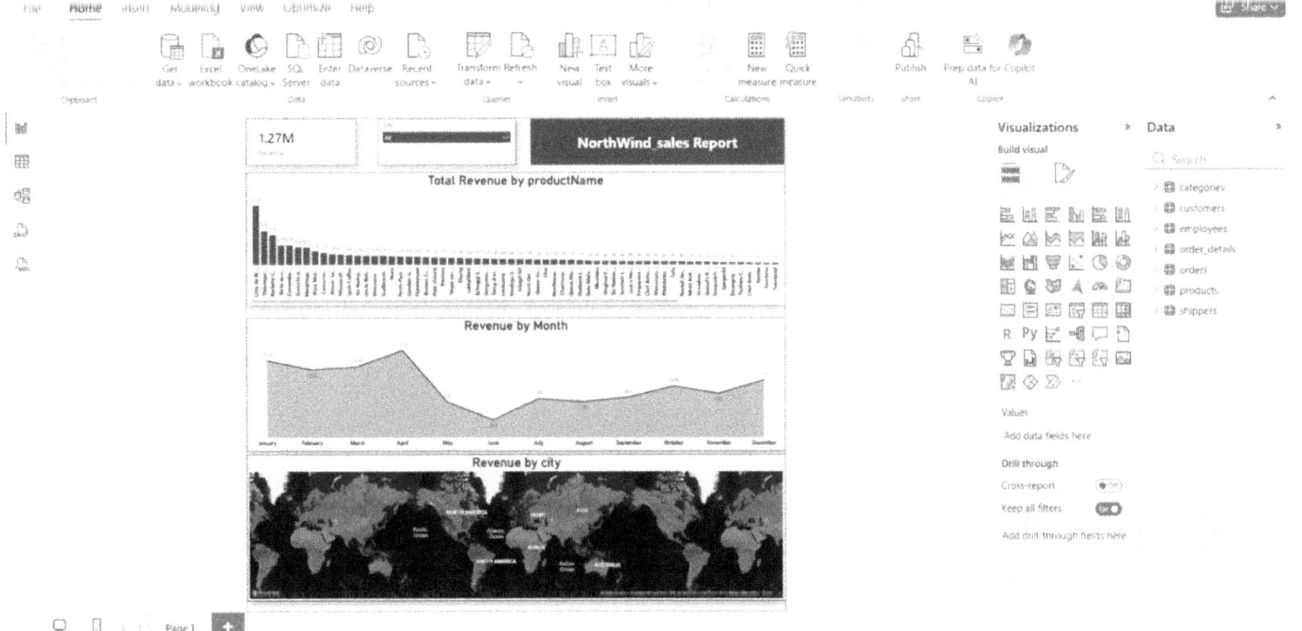

*1.10 - **The Finished Product.** A fully interactive Executive Summary built in under 10 minutes. Notice the layout strategy: The "Big Number" (Card) anchors the view, while the charts break down performance by Geography (Map), Category (Bar), and Time (Line).*

STOP. DON'T TURN THE PAGE YET

CAPTURE YOUR FIRST WIN AND SHARE IT

SNAP A PHOTO OF YOUR DASHBOARD AND SHARE IT USING THE QR CODE TO DOWNLOAD A FREE BONUS

THANKS FOR SHARING

GRAB YOUR BONUS!

SCAN ME!

Day 2 - Data Modeling: Building the Brain of Your Report

2.1 The End of VLOOKUP: Introduction to Relationships

DAY 2

The Paradigm Shift: From "Wide" to "Relational"

You know the sound.

You open a heavy Excel workbook, change a single cell, and suddenly your laptop fan starts screaming like a jet engine on the tarmac. You look at the bottom right corner of your screen, praying for progress, but it just mocks you: *Calculating: (4 Threads) 8%...*

If you are nodding right now, you are suffering from **Spreadsheet Fatigue**.

The culprit? It's not your hardware. It's your **mental model**. In the Excel world, context is king, but it comes at a premium price. If you need to analyze Sales by Region, your muscle memory takes over: you add a column to your Sales table and write a `VLOOKUP` (or `XLOOKUP`) to stamp "North America" onto every single row where a US transaction occurred.

Here is the problem. Technically, you just created a "Wide Table" or Flat File. It feels comfortable because you can see everything in one place. But for your computer's memory (RAM), it is a disaster.

Do the math. If you have 1 million sales rows, you are physically storing the text string "North America Distribution Center" **1 million times**. That is millions of bytes of redundant data clogging the pipes. It's inefficient. It's slow. It's obsolete. **The Power BI Way** demands that you unlearn this behavior immediately. Power BI operates on a **Relational Model** (specifically, the Star Schema). Instead of welding your data into one massive, unwieldy sheet, you keep your tables separate and connect them via "Relationships."

Think of a Relationship as a **Virtual VLOOKUP**. When you build a chart, Power BI looks up the data on the fly, for that specific micro-second, without permanently bloating your file. This leverages the **VertiPaq Engine**, a columnar storage beast that compresses data by storing unique values only once.

By switching from a wide flat file to a relational model, you aren't just organizing data; you are often reducing your file size by 90% and increasing calculation speed by 10x.

The Hard Truth:

You are going to hate this at first. You will feel "blind." Why? Because you can't see the *Customer Name* right next to the *Sales Amount* in the Data View. You will have a frantic urge to merge tables "just to be safe." **Don't do it.** The comfort of seeing one big table is the very thing limiting your ability to scale. Trust the engine.

Core Concepts: The Anatomy of a Relationship

To finally fire your VLOOKUPs, you must understand the two actors in this play. A Star Schema isn't complex theory; it is simply a strict separation of "Nouns" and "Verbs."

1. Fact Tables (The Verbs)

These are your **Actions**. They record what actually happened in the business.

- **Examples:** Sales, Website Visits, Support Tickets, GL Entries.
- **The Profile:** These tables are incredibly long (millions of rows) but usually narrow. They consist almost entirely of numbers (Values) and ID codes (Foreign Keys).
- **The Rule:** You rarely touch these directly. You aggregate them.

2. Dimension Tables (The Nouns)

These are your **Context**. They describe *who, what, where,* and *when*.

- **Examples:** Products, Customers, Employees, Calendar.
- **The Profile:** These tables are short but wide. They contain the descriptive text (Product Name, Color, Size, Manager Name) that you actually want to read on a report.
- **The Rule:** You use these to filter and slice your data.

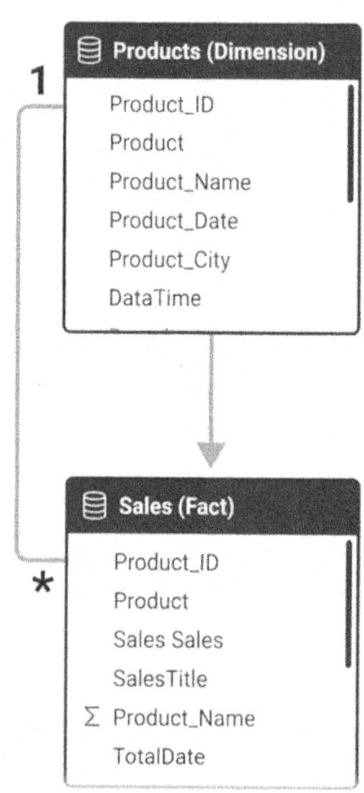

*2.1 - **The Golden Thread:** One-to-Many. This relationship line is the core engine of Power BI. It tells the software: "Every product appears exactly once (1) in the Products table, but can be sold many times (*) in the Sales table." The arrow indicates the direction filters flow—always from the One side down to the Many side.*

The Mechanics of the Wire

To build the bridge, you connect the **Primary Key (PK)** from the Dimension to the **Foreign Key (FK)** in the Fact Table.

- **Primary Key:** The unique identifier in the Dimension (e.g., `Product_SKU`). It must appear **exactly once**. No duplicates allowed. If you have duplicates here, your model breaks.
- **Foreign Key:** The reference in the Fact table. The same `Product_SKU` appears thousands of times (because, hopefully, you sold that product thousands of times).

Filter Propagation: How the Brain Thinks

his is the single most critical concept in Data Modeling. If you get this wrong, your numbers will be wrong. In Excel, VLOOKUP pulls data *into* the table. In Power BI, Relationships push filters *out* of the table.

Imagine a waterfall.

- **The Source (One-Side):** The Dimension Table.
- **The Flow:** The Relationship wire.
- **The Pool (Many-Side):** The Fact Table.

DAY 2

When you click "United States" on a slicer (which comes from your Region Dimension), Power BI doesn't "look up" the US. Instead, it sends a **filter signal** flowing downhill along the relationship wire. It hits the Sales table and instantly hides every row that doesn't carry the US ID code. This is called **Filter Propagation**. Because the engine uses "Dictionary Encoding" (mapping complex text to simple integers in the background), this filtering happens at the speed of RAM. We are talking nanoseconds. Benchmarks suggest this method is **3x to 5x faster** than calculating row-by-row logic in a flat table.

The Danger Zone: Bi-Directional and Many-to-Many

As you drag and drop lines in the Model View, you will eventually encounter two scenarios that look like solutions. They aren't. They are traps.

The "Many-to-Many" (:) Mirage

You try to connect `Sales` directly to `Budget`. Power BI warns you this is a "Many-to-Many" relationship. You click "OK" because the warning goes away and the line appears.

The Reality: You have created chaos. If a Product appears multiple times in Sales and multiple times in Budget, Power BI doesn't know how to match them precisely. The results will be unpredictable.

The Fix: This is almost always a sign of a missing Dimension. You need a unique "Product" table to sit in the middle and filter *both* Sales and Budget.

The "Bi-Directional" (Both) Trap

By default, filters flow downhill (1 to *). You can force them to flow upstream (Both directions). It's tempting. You might want your Sales table to filter your Customer list to show only "Active Customers."

The Cost: Bi-Directional filtering introduces ambiguity. It forces the engine to do double the work and prevents it from "caching" (saving) query plans. In large models, this is the **#1 cause of slow reports**.

The Final Word:

"But it works!" is the most dangerous phrase in data modeling. Just because Power BI *allows* you to create a Many-to-Many relationship or turn on Bi-Directional filtering doesn't mean you should. It's like driving a Ferrari in first gear on the highway; the car moves, but the engine is screaming.

Stick to **One-to-Many (1:*)** and **Single Direction** filters until you are absolutely sure you need to break the rules.

2.2 Star Schema Basics: Fact Tables vs. Dimension Tables

If you have spent years building reports in Excel, you have likely developed a specific survival habit: the giant "Flat File."

You start with a raw list of transactions. Then, you write a dozen `XLOOKUP` or `VLOOKUP` formulas to pull in the Product Name, the Customer Region, the Store Manager, and the Item Category. You drag those formulas down 500,000 rows. Your computer fan starts sounding like a jet engine. You end up with a monolithic table that is 80 columns wide, incredibly sluggish, and constantly on the verge of crashing.

In Excel, combining everything into one massive sheet feels necessary to analyze data. In Power BI, it is a cardinal sin.

Power BI is not a spreadsheet; it operates on a highly optimized database engine. To unlock its speed\u2014to make dashboards cross-filter instantly rather than freezing your screen\u2014you must abandon the flat file. You need to separate the actions from the context. Welcome to the **Star Schema**.

THE FLAT FILE: LEGACY CHAOS

35%

SYSTEM LOAD: HIGH

REDUNDANT DATA

INCONSISTENT

INCONSISTENT

SLUGGISH PERFORMANCE

DATA REDUNDANCY

SCALABILITY ISSUES

DIFFICULT TO MAINTAIN

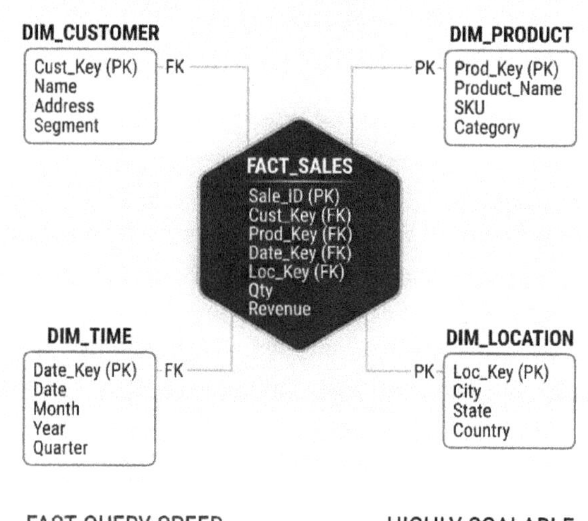

*2.2 - **The Paradigm Shift.** In Excel, combining everything into a massive "Flat File" (Left) feels necessary, but it forces the engine to process millions of redundant text strings, resulting in sluggish, freezing reports. The Power BI standard is the Star Schema (Right): an elegant, lightweight architecture where facts sit in the center and descriptive context surrounds it, unlocking instantaneous calculation speeds.*

Fact Tables: The Verbs (What Happened?)

Think of a Fact table as the ledger of actions. It records events. Someone clicked a link, a shipment left the warehouse, a payment processed.

Let's look at our case study, SmartGear Retail. Every time a customer buys something, a record is created. This goes into the `Fact_Sales` table.

The golden rule of a Fact table is strict minimalism. It should only contain two things: numbers you can do math on (like `Quantity`, `Discount`, or `Revenue`) and IDs (like `ProductID` or `StoreID`).

It is built to be **long and narrow**.

It will grow by tens of thousands of rows a day, but it should only be a few columns wide. You will never find the descriptive text "SmartGear Pro-Gaming Laptop" here. You will only find `ProductID: 402`. Fact tables are lean, pure, mathematical machines.

Dimension Tables: The "Nouns" (Who, Where, When?)

DAY 2

If the Fact Table is the transaction, the Dimension Tables are the dictionaries that make sense of it. They provide the context. They answer the "W" questions: *Who* bought it? *Where* was it sold? *When* did it happen?

This is the **"Human Interface"** of your data model. When you build a report and drag a field into a Slicer, a Legend, or an Axis of a bar chart, you are almost always touching a Dimension Table. The CEO looking at your dashboard doesn't interact with the raw numbers; they interact with the descriptions. These tables feel stable, readable, and comforting.

The Logic (The Remote Control): Dimension tables are designed strictly for filtering and grouping. Think of a Dimension table as the remote control, and the Fact table as the television. When you click "Audio Equipment" on your remote (the Dimension Slicer), it sends an invisible signal down the relationship wire. It instantly tells the massive television (the Fact table) to hide every channel (row) that doesn't match "Audio Equipment." Because Dimension tables are tiny—often just a few thousand rows compared to millions of sales—the Power BI engine can scan them in a fraction of a millisecond. It is the secret to instantaneous dashboards.

The Hard Truth (The UI Trap): Beginners often neglect Dimension tables, treating them like backend storage and leaving them messy. This is a fatal UX (User Experience) mistake. In Excel, a messy column header like Cust_Nm_V2_FINAL doesn't matter as long as your VLOOKUP works. In Power BI, **your column names are your user interface.** Since Dimension tables provide the fields your end-users actually see in the filter pane and on the charts, a sloppy naming convention makes your entire report look unprofessional, regardless of how accurate the underlying math is. You must rename these columns in Power Query to plain, readable English (e.g., Customer Name).

Characteristics of a Dimension Table:

Short and Wide: Embrace the width. While Fact tables must be kept razor-thin to save RAM, Dimension tables are allowed to be generously wide. SmartGear Retail might have 10 million sales rows (Fact), but only 500 unique products in its catalog (Dimension). Because 500 rows consume almost zero memory, you can afford to have 30 columns describing every aspect of those products: Category, Sub-Category, Color, Weight, Manufacturer, Launch Date. More columns here mean more ways for your users to slice and analyze the data later.

Low Volatility: Data here changes infrequently. A customer moves to a new city, or a new product line launches, but these are slow-moving, administrative updates. They are not the high-velocity, second-by-second events of a cash register.

Text Heavy: This is where the strings live. Names, Categories, Colors, Regions. Text is the heaviest data type for a computer processor to read. By isolating all this heavy text inside a small, separate Dimension table, we prevent the Power BI engine from having to read the word "Smartphone" ten million times in the Fact table. It only has to read it once.

The Golden Rule (The Primary Key): This is non-negotiable. Every Dimension table **must** contain one column where every single value is 100% unique. We call this the Primary Key (e.g., Product_ID). In Excel, a duplicate row is a minor annoyance. In a Power BI Dimension table, a duplicate Product_ID is fatal. If the engine looks up an ID and finds two different products, it panics and breaks the relationship. Your Dimension table must be an absolute, unique list of entities.

DAY 2

The "Star" Topology: Why the Engine Loves It

When you arrange these tables in the Model View, you place the Fact Table in the center and the Dimension Tables surrounding it. The relationships radiate outward. Like a star. Why does this matter? Physics. Power BI runs on an in-memory engine called **VertiPaq**. This engine is columnar, meaning it compresses columns individually.

- If you have a Fact Table with 10 million rows, and you store the text "United States" in every row, the engine has to work incredibly hard to store and scan that text.
- In a Star Schema, the engine stores the number "1" in the Fact Table (representing the Country ID). It stores "United States" **once** in the Dimension Table.

When you filter for "United States," the engine grabs the ID "1" from the small Dimension table and instantly filters the 10 million rows in the Fact table using simple integer matching. This is exponentially faster than text matching.

Summary: The Cheat Sheet

The most common reason for a slow Power BI report is not "too much data." It is a model that failed to distinguish between Facts and Dimensions.

Feature	Fact Table	Dimension Table
Role	The Event (Measurement)	The Context (Description)
Grammar Analogy	**Verbs** (Sold, Clicked, Shipped)	**Nouns** (Customer, Product, Date)
Key Type	Foreign Keys (FK) - Many duplicates	Primary Keys (PK) - Unique values
Content	Mostly Numeric (IDs & Amounts)	Mostly Text & Dates
Shape	**Long** (High Row Count)	**Wide** (High Column Count)
Power BI Job	To be **Summarized**	To be **Filtered**
Example	Sales_Transactions	Dim_Customers

2.3 Cardinality Explained: One-to-Many vs. Many-to-Many

Cardinality Explained: One-to-Many vs. Many-to-Many

Cardinality is often taught as abstract database theory. Ignore that. For you, the analyst, cardinality is simply **the physics of your report.** Think about your life in Excel. When you execute a `VLOOKUP` or `XLOOKUP`, you are hunting for a specific match. You find it, you grab the value, you stop. Done. Power BI operates differently. We aren't just looking up isolated values; we are engineering how filters—context—flow through the entire model. Cardinality answers a critical structural question: **"How unique is the data in this column, and what are the rules of engagement for filters traveling through it?"**

DAY 2

Get this wrong, and your report becomes a sluggish mess that gives you plausible, yet mathematically incorrect, numbers. Get it right, and the model feels snappy, intelligent, and robust.

1. The Gold Standard: One-to-Many (1:*)

This relationship is the heartbeat of the **Star Schema**. It creates a rigid, predictable, and highly efficient highway for your data. It connects a **Dimension** (the "One" side, representing the *noun*) to a **Fact** (the "Many" side, representing the *verb*).

- **The "One" Side (The Anchor):** This is your Dimension Table. It holds the definition of truth—the unique ID. In `Dim_Products`, the Product ID `P-101` appears exactly once. No duplicates. No ambiguity.
- **The "Many" Side (The Event):** This is your Fact Table. In `Fact_Sales`, that same Product ID `P-101` might appear thousands of times—once for every transaction involving that product.

The Physics of the Waterfall

Visualize a One-to-Many relationship as a cascading waterfall. Gravity—your filter logic—naturally pulls water from the top down. When you click "Red" in your Product table (the top of the waterfall), that selection flows effortlessly down the relationship line. It hits the Sales table and instantly isolates only the rows that match "Red." It is fast because the engine doesn't have to guess. The path is mathematically singular. Gravity does the work.

2. The Truth About Many-to-Many (*:*)

Here is the brutal reality: **Many-to-Many relationships are usually a mistake.** In the Excel world, you can get away with fuzzy matching or manual overrides. in a data model, ambiguity is fatal. A Many-to-Many relationship occurs when a specific value appears multiple times in *both* columns you are trying to connect.

The Scenario

Imagine you have a `Sales_Team` table and a `Regions` table.

- Salesperson "Sarah" works in both the "North" and "East" regions (Sarah appears twice).
- The "North" region is covered by both "Sarah" and "Mike" (North appears twice).

The Engine's Struggle

If you filter for the "North" region, which salespeople should the report show? Just Mike? Mike and Sarah? And if you show Sarah, should the report also calculate her sales from the "East" region, since she is the same person? To solve this, Power BI is forced to perform "Bi-Directional Filtering." Instead of a waterfall, you have a two-way street with no traffic lights. Filters bounce back and forth, creating an echo chamber.

Why Experienced Analysts Avoid It:

1. **Performance Tax:** The engine cannot use its optimized join algorithms. It must create temporary internal tables to map the chaos on the fly. This consumes memory and creates noticeable lag in your visuals.

2. **The "Ambiguity" Trap:** If you have multiple Many-to-Many relationships in one model, you risk **circular dependency**. The engine literally doesn't know which path to take to calculate a total. It might disable a relationship entirely or—worse—give you a number that looks right but is mathematically wrong.

3. The Professional Fix: The Bridge Table

You are a knowledge worker looking for high efficiency and low friction. While it seems "easier" to just drag a line between two messy tables and let Power BI figure it out, that is high friction in the long run. It is sloppy architecture.

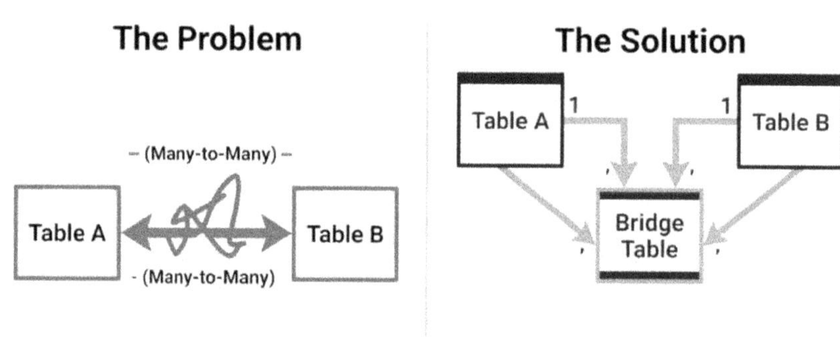

*2.3 - **The Bridge Table Architecture.** A direct Many-to-Many relationship (Left) creates an ambiguous loop where filters bounce back and forth unpredictably. The professional solution (Right) is to insert a "Bridge Table" of unique IDs. This breaks the chaotic loop into two clean, one-directional One-to-Many relationships flowing downward like waterfalls.* The professional move is to resolve the conflict using a **Bridge Table** (also known as a Junction Table).

How it Works

Don't force the two messy tables to talk directly. Introduce a neutral third party—a mediator.

1. Identify the unique values from both sides (e.g., a list of every unique Employee/Region assignment).

2. Create a new table that sits in the middle.

3. Connect both original tables to this new table.

The Result

You transform one chaotic `:` relationship into two clean, stable `1:*` relationships. The Bridge Table becomes the "Many" side for both dimensions. This allows filters to flow down from your dimensions into the bridge, meeting in the middle without ambiguity.

The Architect's Advice:

If you find yourself tempted to use Many-to-Many because you are in a rush, stop. It is technical debt. Building a Bridge Table takes five minutes; debugging a report that totals sales incorrectly because of bi-directional filtering can take days. Always choose the clean waterfall over the chaotic two-way street.

2.4 The Calendar Table: The Heart of Time Intelligence

The Mindset Shift: From "Cell Formatting" to "Central Clock"

Listen. If you take only one thing away from this entire book, make it this. Skip this section, and your fancy "Year-over-Year" growth charts? Dead. Your Moving Averages? Broken. In the Excel world, a date is just a formatted cell. In the Power BI world, Time is a ruthless mathematical dimension. It requires strict, unbroken continuity.

The Why. In Excel, time is passive. You make a PivotTable, group by month, and if you had zero sales in April, April simply vanishes from the report. You probably think that's a feature. It isn't. In Power BI, that "gap" is fatal. The **Time Intelligence** engine runs on calculus-like continuity. When you ask for a "Previous Day" calculation on a Monday, the engine mathematically hunts for the predecessor ($Monday - 1 = Sunday$). If your sales data has no rows for Sunday because the office was closed, and you rely solely on your sales table for dates, the engine hits a void. It crashes. It returns a blank. The coordinate it was looking for literally does not exist in your model.

The Experience. Think of your data model as a high-performance orchestra. Your Sales table is the strings; your Budget table is the brass section. The **Calendar Table** is the Conductor. Without a conductor, the strings play at their own speed (dates where you had sales), and the brass plays at a different speed (dates where you have budget goals). Chaos. The Calendar Table provides the "Central Clock"—a steady, unbroken beat that forces every other table to align to the same timeline. It ensures that when you say "Q1," the Sales table and the Budget table both know exactly when the music starts and stops.

The Reality Check. You cannot "VLOOKUP" your way out of this. Beginners try to hack this by creating helper columns in their Sales table called "Month" or "Year." This is a trap. If you rely on dates inside your Sales table, you can never report on a day where no sale occurred. You want to show your boss a report with "0 Sales" on a holiday? You can't. That row doesn't exist. You need a separate table to tell Power BI that the day existed, even if the revenue didn't.

DAY 2

The Trap: "Auto Date/Time" (The Performance Vampire)

By default, Power BI tries to mimic Excel's ease of use. It activates a setting called **"Auto Date/Time."** This is a problem.

The Why. When this setting is on, Power BI scans every single column formatted as a Date in your entire model and secretly generates a hidden table for it, complete with hierarchies for Year, Quarter, Month, and Day.

The Experience. This feature is a "Performance Vampire." It looks helpful on the surface. In reality, it is draining the lifeblood of your report. Imagine you have a Sales table with 10 million rows and three date columns: *Order Date*, *Ship Date*, and *Due Date*. If "Auto Date/Time" is on, Power BI silently creates three distinct, hidden tables behind the scenes. Your file size explodes. Your refresh times crawl. You are hiking up a mountain carrying a backpack full of rocks that you didn't pack.

The Solution. Turn it off. Now. Go to *File > Options > Data Load* and uncheck **"Auto Date/Time"**. Microsoft leaves this on to help casual users dragging and dropping simple lists. But you are building a professional model. It handles standard years (Jan-Dec) fine, but the moment your CFO asks for a "Fiscal Year" starting in April, the Auto Date/Time logic collapses. It creates a bicycle with training wheels welded on; you cannot ride it in the Tour de France.

The Architecture: One Calendar to Rule Them All

The Calendar Table serves a critical architectural role in the **Star Schema**. It acts as a "Shared Dimension." Consider your actual business environment. You likely have "Actuals" (Sales table) and "Targets" (Budget table).

- Sales are recorded daily.
- Budgets are often recorded monthly.
- They live in two different Fact tables.

Without a common Calendar table, you cannot plot these two on the same chart. You cannot filter both by "2023" simultaneously. The Calendar table sits above them, filtering both flows of data through a single, unified lens.

The 5 Golden Rules of a Valid Date Table

To function as the "Central Clock," your Calendar table must obey the laws of the DAX engine. These are not suggestions; they are requirements.

1. **Contiguous Dates:** It must contain *every single day* between the start and end date. No skipped weekends. No skipped holidays. Time is a continuum, not a collection of workdays.

2. **Unique Values:** Each date appears exactly once. It is the primary key.

3. **No Blanks:** The date column strictly cannot contain null values.

4. **Full Years:** It is best practice to cover complete years (Jan 1 to Dec 31). If your sales start in March, your Calendar should still start Jan 1. This ensures year-end calculations work correctly.

5. **Mark as Date Table:** You must explicitly register this table with the engine.

The "Mark as Date Table" Switch

This is the handshake. It is the moment you introduce your custom table to the Power BI Time Intelligence brain.

Action: Right-click your Calendar table > *Mark as date table.*

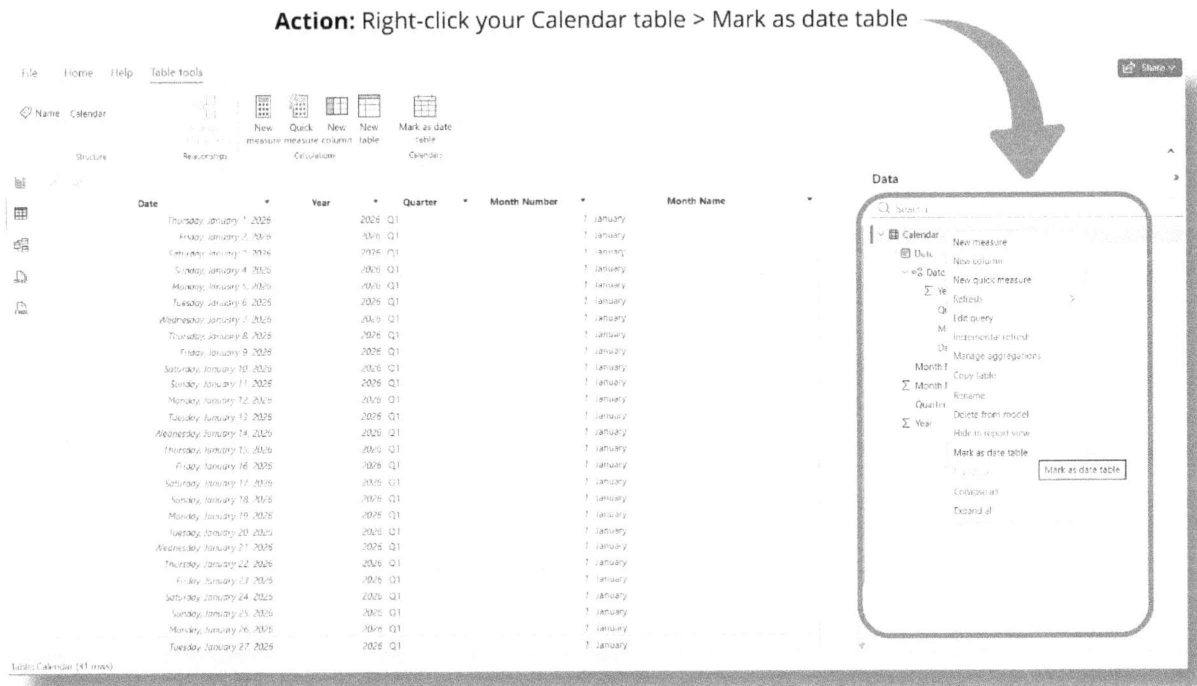

*2.4 - **The Time Intelligence Handshake.** Simply naming a table "Calendar" is not enough. You must explicitly introduce it to Power BI's engine by right-clicking and selecting "Mark as date table". The green validation checkmark confirms your dates are contiguous and unique—the strict prerequisite for functions like.*

The Why. When you do this, two technical shifts occur:

1. **Cleanup:** Power BI stops trying to build internal hierarchies for this table.

2. **Filter Context Protection:** It tells DAX, "When I calculate data based on dates, use *this* specific column as the anchor." This fixes a common bug where filtering by "Month" fails to sort the days correctly unless this setting is active.

Implementation: DAX vs. Power Query

You will see tutorials telling you to write a quick DAX formula (` CALENDARAUTO()`) to build this table. Don't rush into that.

- **DAX (Calculated Table):** It's fast. It's easy. It scans your data and builds the table instantly. *However*, it is calculated at runtime.
- **Power Query (M Code):** This is the professional choice.

The Recommendation.

If you are just prototyping, use DAX (`CALENDARAUTO`). It's quick and dirty. But if you are building a report that needs to handle your company's specific "4-4-5 Fiscal Calendar" or exclude bank holidays from a "Days to Ship" calculation, you must build this in **Power Query**. Why? Customization. You will eventually need columns like "Fiscal Month," "Working Day vs. Weekend," or "Company Holiday." Do the heavy lifting during the data load (Power Query), not during the user interaction (DAX).

DAY 2

The Flow. You have now moved from thinking of dates as labels on a chart to understanding Time as the central spine of your data model. By turning off "Auto Date/Time" and building a dedicated Calendar, you have removed the training wheels. You are ready to control time, rather than letting it control you. Now, let's connect this spine to the rest of the body.

2.5 Creating Hierarchies for Drill-Down Capabilities

So, you killed "Auto Date/Time." Good. But you might have noticed an immediate side effect: your dates no longer automatically cascade from Year to Quarter to Month. The automatic dropdowns are gone. Do not panic. This is intentional. You have effectively traded a "black box" feature—convenient but opaque—for absolute, granular control. You must now explicitly tell Power BI which columns belong together. In the Excel world, "drilling down" is often a clumsy affair.

You double-click a PivotTable cell, and suddenly you are staring at a new sheet full of raw data you didn't ask for. Or, you manually stack five different fields into the "Rows" area, creating a stair-stepped, cluttered nightmare. Power BI handles this differently. We use **Hierarchies**. Think of a Hierarchy not just as a group of columns, but as a pre-designed user interface for your future self. It is a logical path of analysis. When you drag a Hierarchy onto a visual, the granular details (like City, SKU, or Daily Sales) are hidden in the background, waiting for permission to appear. You keep the canvas clean until the user asks for more depth.

Common "Analysis Paths" you should build immediately:

- **Geography:** Country > State > City > Zip Code.
- **Product:** Category > Sub-Category > SKU.
- **Time (The Standard):** Year > Quarter > Month > Date.

Implementation: The "Context Menu" Method

You *can* build hierarchies by dragging and dropping fields on top of each other in the Model View. **Don't.** The mouse interface is finicky. Accidentally dropping a "Month" column into a "Product" folder happens more often than you would think, and undoing it disrupts your flow. For high-efficiency modeling, we rely on precision. Use the Data pane context menus.

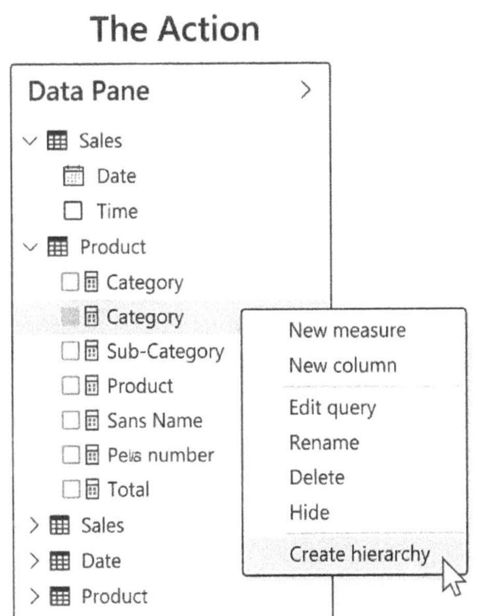

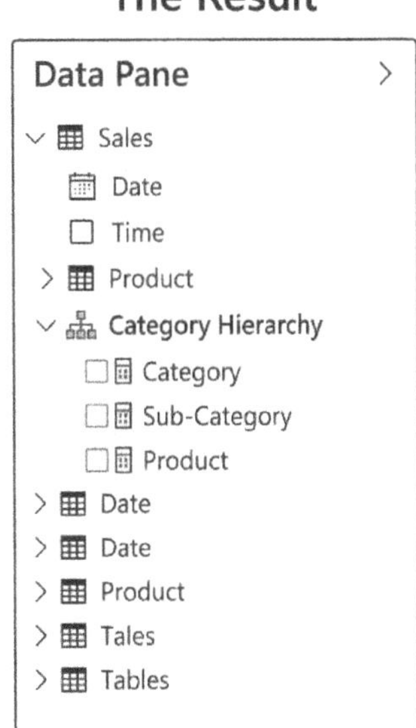

*2.5 - **The Power of Pre-Built Paths.** Stop dragging three separate fields into every single chart you build. By right-clicking your top-level category and selecting "Create hierarchy" (Left), you bundle them into a single, logical unit (Right). From now on, dragging just the "Hierarchy" icon into a visual instantly enables Drill-Down capabilities.*

DAY 2

1. **Identify the Parent:** Find your highest-level field (e.g., `Category`).

2. **Initiate:** Right-click the field and select **Create hierarchy**. Power BI will generate a new icon that resembles an organizational tree chart.

3. **Add Children:** Right-click the next level down (e.g., `Sub-Category`), select **Add to hierarchy**, and choose the parent you just created.

4. **Verify & Order:** Expand the new Hierarchy icon. Logic is paramount here: Year must sit above Month. If the order is wrong, drag the items within this specific list to rearrange them.

The Truth About "Drill" Buttons

Here is the reality that trips up almost every Excel user transitioning to Power BI: **Visuals behave differently depending on how you click them.** Once you place a Hierarchy on a visual (like the X-Axis of a bar chart), Power BI activates the **Drill Header**—a cluster of arrows at the top of the chart. If you do not understand the distinction between the "Double Arrow" and the "Pitchfork," you will present misleading data. It is that simple. Let's break down the three distinct interactions.

1. The Microscope (Single Down Arrow)

This is **Drill Mode**. When you toggle this button on (it turns dark), you are changing your mouse cursor into a filter. Clicking a specific bar on the chart (e.g., "2023") filters the entire visual to show *only* the next level for that specific item.

- *The Insight:* "I want to investigate exactly why 2023 was so profitable."

2. The Aggregator (Double Down Arrow)

Warning: This is the button that causes the most confusion.

This button ignores the parent context entirely. If you are viewing Years and click this, Power BI jumps to the next level (Quarter) but aggregates **ALL** years together. It will show you a single bar for "Q1." That bar is not Q1 2023. It is the sum of 2023 Q1 + 2024 Q1 + 2025 Q1.

- *The Trap:* Users click this expecting a chronological timeline and get confused by weirdly high numbers.
- *The Real Use Case:* **Seasonality Analysis.** It answers the question, "Do we generally sell more in Q4 than Q1, regardless of the year?"

3. The Storyteller (Pitchfork / Split Arrow)

This is the button you likely want 90% of the time.

It creates a **Continuous Flow**. It expands the hierarchy down to the next level while preserving the parent label. It displays "2023 Q1, 2023 Q2... 2024 Q1, 2024 Q2." It respects the timeline.

- *The Insight:* **Trend Analysis.** It shows the full history without losing the year context.

Day 3 - DAX Fundamentals: Speaking the Language of Data

3.1 DAX vs. Excel Formulas: Understanding Evaluation Context

The Core Paradigm Shift: From Coordinates to Context

If you are an Excel veteran, your brain operates on a grid. You have been wired to think in **spatial coordinates**. When you type `=A2+B2`, you are essentially playing a game of *Battleship*: you tell the software exactly *where* the data lives. It is a rigid, reliable map. You point to cell C2, and you say, "The answer is here." DAX obliterates this map. In the Data Model, there is no "cell C2." There is no "row 40." There are only **Tables** and **Columns**. When you write `Total Sales = SUM(Sales[Amount])`, there is no coordinate telling the engine which rows to add up.

DAY 3

The Spreadsheet Mindset - Red/Grey Theme

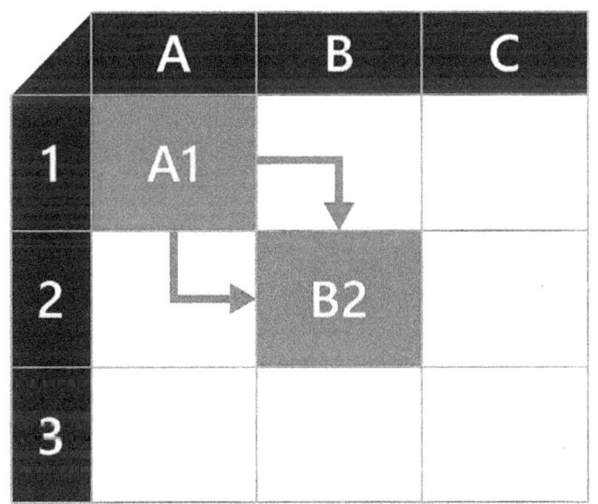

The Database Mindset - Navy/Teal Theme

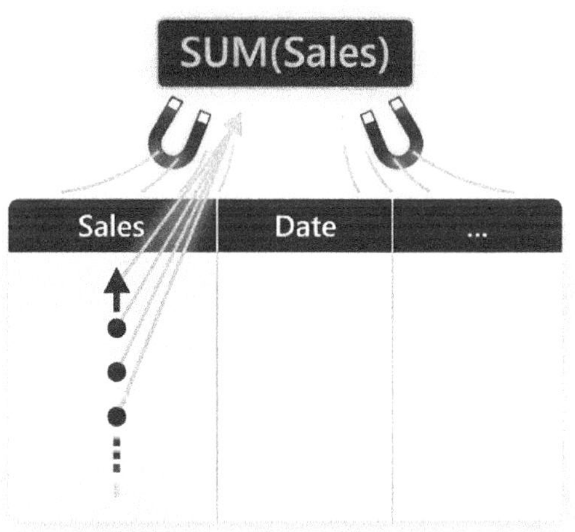

*3.1 - **The Mental Shift.** The hardest part of learning DAX is unlearning Excel. In Excel (Left), you navigate a rigid grid using coordinates (e.g., A1+B2). In Power BI (Right), cells do not exist. You are writing logic for entire columns (e.g., SUM(Sales)), which dynamically aggregate based on the filters applied by the user.*

Why would anyone voluntarily give up that level of control? **Scalability.** Excel's cell-based logic is computationally expensive to maintain. If you have a million rows, Excel essentially tracks a million individual dependencies. It drags. It crashes. DAX, however, utilizes a columnar database engine (VertiPaq). It compresses data by grouping values in columns, not cells.

To perform a calculation, it doesn't look for a "place"; it looks for a "state of being." Here is the uncomfortable reality: **Your spatial intuition is your enemy right now.** In Excel, you see the data, then you write the formula. In DAX, you write the formula, and it creates the result based on an invisible set of rules called the **Evaluation Context**. Think of this context as the engine constantly asking: *"Given the current environment—the filters, the slicers, the row headers—what subset of data is actually visible to this formula right now?"*

The Two Types of Evaluation Context

To master this, you must understand that the engine is always running two distinct processes depending on *how* you are calculating the data. This distinction is the single biggest technical hurdle for professionals moving from spreadsheets to Power BI.

A. Row Context: "The Tunnel Vision"

Imagine you are auditing a printed spreadsheet with a physical ruler in your hand.

You place the ruler under the first row. You read the `Price` on that specific line, multiply it by the `Quantity` on that same line, and write the result. Then, you slide the ruler down. You do not care what is happening ten rows down or ten rows up. You are hyper-focused on the current line. This is **Row Context**. It is the engine's ability to distinguish the "current row" from all others during an iteration.

3.2 - ***Understanding Row Context.*** *When DAX uses Row Context (like in a Calculated Column or an Iterator function like.*

Where it lives:

- **Calculated Columns:** When you add a hard-coded column to a table, DAX automatically creates a Row Context to calculate the value for every single row, one by one.
- **Iterators (The "X" Functions):** Functions like `SUMX` or `FILTER` force the engine to generate this "ruler" behavior, stepping through a table line by line to perform a calculation.

But here is the trap. Do not mistake Row Context for a filter. If you are in a Row Context (iterating through the *Sales* table) and you try to sum up the *Sales* column, DAX will simply sum the **entire column**, ignoring the row you are currently standing on. Why? Because Row Context tells DAX *which row you are on*, but it does not inherently *filter* the view of the data. It gives you the values of the current row. Nothing more.

B. Filter Context: "The Force Field"

Imagine a giant funnel suspended above your data table. Before any math happens—before a single number is summed or averaged—the user interacts with your report.

- They click a Slicer for "Year: 2024." -> The funnel slams shut on all 2023 data.
- They look at a bar chart for "Audio" products. -> The funnel blocks all non-Audio data.

What falls through the funnel to reach your formula is the **Filter Context**. Unlike Excel formulas, which are static (cell C2 always equals A2+B2), a DAX measure is a chameleon. It changes its value based on the environment. The measure `[Total Sales]` changes its output depending on the slicers and visuals surrounding it.

DAY 3

Where it comes from:

- **Slicers & Page Filters:** The obvious user inputs.
- **Visual Coordinates:** This is the subtle one. In a Matrix (Pivot Table), the row header "North" is not just a label; it is a filter command. It effectively tells the measure: "Calculate yourself, but only for the North region."

In Excel, you filter data *after* you see it. In DAX, the Filter Context filters the data *before* the formula ever runs. If your numbers look wrong, 99% of the time it is not your math; it is a hidden filter (a "rogue force field") you forgot about.

Practical Application: How DAX "Thinks"

Let's slow down time. We are going to watch exactly what happens when DAX renders a Matrix visual showing **Total Sales** by **Region.**

The Scenario: You have a measure `Total Sales = SUM(Sales[Amount])`. You drop it into a matrix with "Region" on the rows. You see a number for the "North" region.

The Invisible Workflow:

1. **The Trigger:** The Matrix visual requests a value for the cell corresponding to the row "North".
2. **Setting the Scene (Filter Context):** DAX pauses. It assesses the environment.
   - Is there a slicer for Year? Yes, 2024. -> *Filter 1 applied.*
   - What is the row header? "North". -> *Filter 2 applied.*
   - DAX goes to the original *Sales* table and hides every single row that is NOT "North" and NOT "2024".

3. **The Calculation:** Now, and only now, does DAX look at the formula: `SUM(Sales[Amount])`. It looks at the *Sales* table. It sees only the rows that survived the filters (the "North/2024" rows). It sums them up.

4. **The Output:** The result is returned to the Matrix cell.

DAX FILTER CONTEXT SEQUENCE

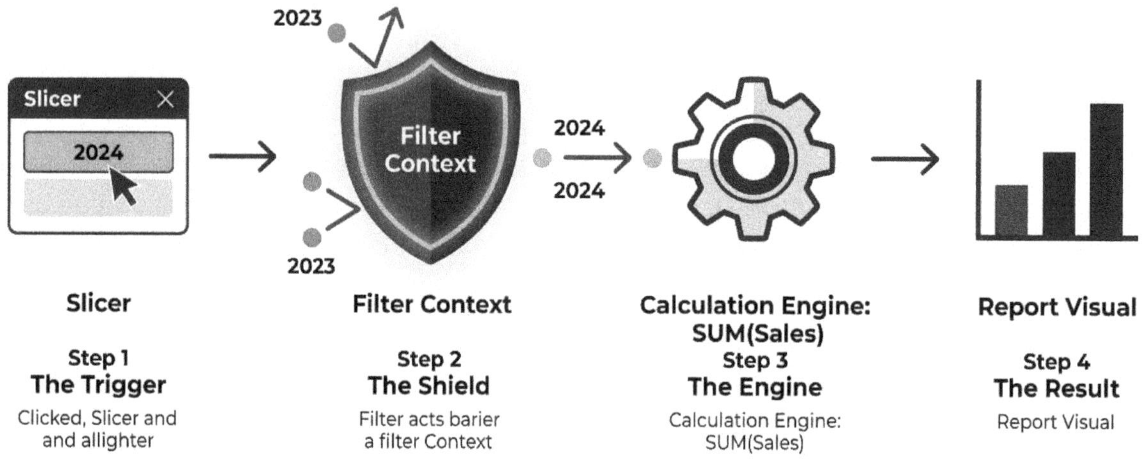

*3.3 - **The Anatomy of Filter Context.** This is the secret to DAX: The math happens last. When a user clicks "2024" (Step 1), a shield forms around the data model (Step 2), blocking all other years. The engine (Step 3) only ever sees the surviving data droplets, aggregating them for the final visual (Step 4).*

In Excel, the cell contains the value. In Power BI, the cell contains a **query**. Every time you click a slicer or scroll a chart, you are not just moving a picture; you are firing off a fresh set of queries to the engine, asking it to reconstruct reality based on a new Context. Understanding this flow—Filter first, Calculation second—is how you stop guessing and start engineering.

3.2 Calculated Columns vs. Measures: When to Use Which

The Fundamental Shift: Static vs. Dynamic

For an Excel professional, the instinct to "add a column" is muscular memory. If you need `Price * Quantity`, you create a new column. If you need to extract the year from a date, you create a new column. In the world of DAX, blindly following this instinct is the fastest way to bloat your model and kill performance. Understanding the distinction between **Calculated Columns** and **Measures** is not just a syntax detail; it is the architectural foundation of a scalable Power BI report.

Calculated Columns: The "Hard-Coded" Data

Think of a Calculated Column as physically stamping new data into your table. When you create one, DAX runs the calculation row-by-row immediately, stores the result in your computer's RAM, and saves it as part of the file.

Because they are computed during the data refresh—not when you interact with the report—they are static. They do not change when a user clicks a slicer or filters a visual.

When to use a Calculated Column:

Use them *only* when you need to use the result in a **Filter**, **Slicer**, or as an **Axis** (Row/Column) in a visual.

- **Scenario:** You want to slice your sales report by "High Margin" vs. "Low Margin" transactions.
- **Action:** Create a Calculated Column to tag each row. You can now drag this field into a Slicer.

Measures: The Dynamic Engine

Measures are where the true power of DAX resides. A Measure is not stored data; it is a formula waiting to be executed. It exists only in metadata. The calculation runs on the CPU at the exact moment a user interacts with the report, recalculating instantly based on the current context (filters, slicers, selection).

When to use a Measure:

Use them for numerical values that need to be aggregated, calculated, or displayed in the **Values** area of a visual.

- **Scenario:** You want to see the *weighted average margin* for the selected region.
- **Action:** Create a Measure. If the user selects "Europe," the measure calculates for Europe. If they select "Asia," it recalculates for Asia instantly.

The Golden Rule of DAX Efficiency

To keep your reports fast and your file size small, adopt this simple heuristic:

If you can do it with a Measure, do it with a Measure.

Resort to Calculated Columns only when the specific result is required for grouping or filtering data. If it is a number meant to be analyzed (summed, averaged, counted), it belongs in a Measure.

The Truth About Aggregation: Why "Helper Columns" Fail

In Excel, creating a "helper column" is muscle memory. If you want to know the profit margin, you create a new column, divide Profit by Sales, and drag the formula down. It feels safe. It feels visible. In Power BI, that same habit is a silent killer. Let's look at a specific scenario at **SmartGear Retail** that trips up almost every new analyst. We are analyzing two distinct transactions:

1. **The Enterprise Deal:** We sell a bulk shipment of high-end Laptops. The revenue is massive ($2,000), but because of corporate discounts, the profit is razor-thin (## Calculated Columns vs. Measures: When to Use Which

If there is a single concept that distinguishes a true Power BI professional from an Excel user merely "getting by," this is it.

DAY 3

In Excel, equality rules. Every cell is the same. You type a number, it sits there. You write a formula, it sits there. You can see it, touch it, and audit it. Power BI forces you to abandon this comfort zone. It introduces a divergence that feels counterintuitive, perhaps even wrong at first: **Calculated Columns** versus **Measures**.

They both utilize DAX. They appear identical in the formula bar. But under the hood? They behave like matter and energy. Confusing the two isn't just a syntax error; it is the primary reason your reports lag, your file size bloats, and your numbers turn out mathematically impossible.

00).

2. **The Accessory Add-on:** We sell a single premium USB-C cable. The revenue is negligible ($20), but the markup is enormous, yielding an ## Calculated Columns vs. Measures: When to Use Which

If there is a single concept that distinguishes a true Power BI professional from an Excel user merely "getting by," this is it. In Excel, equality rules. Every cell is the same. You type a number, it sits there. You write a formula, it sits there. You can see it, touch it, and audit it. Power BI forces you to abandon this comfort zone. It introduces a divergence that feels counterintuitive, perhaps even wrong at first: **Calculated Columns** versus **Measures**.They both utilize DAX.

They appear identical in the formula bar. But under the hood? They behave like matter and energy. Confusing the two isn't just a syntax error; it is the primary reason your reports lag, your file size bloats, and your numbers turn out mathematically impossible.

8 profit.

If you use a **Calculated Column** to determine your margin (`Row Profit / Row Sales`), the database calculates the percentage for every single row, individually, the moment the data loads.

If there is a single concept that distinguishes a true Power BI professional from an Excel user merely "getting by," this is it.

In Excel, equality rules. Every cell is the same. You type a number, it sits there. You write a formula, it sits there. You can see it, touch it, and audit it. Power BI forces you to abandon this comfort zone. It introduces a divergence that feels counterintuitive, perhaps even wrong at first: **Calculated Columns** versus **Measures**.

They both utilize DAX. They appear identical in the formula bar. But under the hood? They behave like matter and energy. Confusing the two isn't just a syntax error; it is the primary reason your reports lag, your file size bloats, and your numbers turn out mathematically impossible.

00, Margin 5%. Row 2 (Cable): Sales $20, Profit ## Calculated Columns vs. Measures: When to Use Which

If there is a single concept that distinguishes a true Power BI professional from an Excel user merely "getting by," this is it.

In Excel, equality rules. Every cell is the same. You type a number, it sits there. You write a formula, it sits there. You can see it, touch it, and audit it. Power BI forces you to abandon this comfort zone. It introduces a divergence that feels counterintuitive, perhaps even wrong at first: **Calculated Columns** versus **Measures**. They both utilize DAX. They appear identical in the formula bar.

But under the hood? They behave like matter and energy. Confusing the two isn't just a syntax error; it is the primary reason your reports lag, your file size bloats, and your numbers turn out mathematically impossible.

8, Margin 90%.

Now, you drag that Margin column into a report to show the CEO our overall performance. Power BI does exactly what it is designed to do with columns: **it aggregates them.** usually by averaging.

Common Mathematical Error in Corporate Reports: Averaging Percentages Improperly.

The Trap: Calculating a straight average of two distinct percentage margins is mathematically incorrect.

Product	Sales	Profit	Margin % (Column)
Laptop	$2000	$100	5%
USB Cable	$20	$18	90%
Total/Average		(5% + 90%) / 2 = 47.5% ✗	

*3.4 - **The "Average of Averages" Trap.** If you calculate margin row-by-row using a Calculated Column, Power BI will simply average the resulting percentages (5% and 90%). The report claims you have a massive 47.5% margin, when in reality your total profit ($118) divided by total sales ($2020) is less than 6%. Presenting this false number to a CEO is a career risk.*

It takes the Laptop's **5%** and the Cable's **90%**, adds them up, and divides by two.

The Report Says: 47.5% Margin.

The Reality: You sold $2,020 worth of goods and made ## Calculated Columns vs. Measures: When to Use Which.

If there is a single concept that distinguishes a true Power BI professional from an Excel user merely "getting by," this is it.

In Excel, equality rules. Every cell is the same. You type a number, it sits there. You write a formula, it sits there. You can see it, touch it, and audit it. Power BI forces you to abandon this comfort zone. It introduces a divergence that feels counterintuitive, perhaps even wrong at first: **Calculated Columns** versus **Measures**.

They both utilize DAX. They appear identical in the formula bar. But under the hood? They behave like matter and energy. Confusing the two isn't just a syntax error; it is the primary reason your reports lag, your file size bloats, and your numbers turn out mathematically impossible.

18 in total profit. Your actual margin is ## Calculated Columns vs. Measures: When to Use Which

If there is a single concept that distinguishes a true Power BI professional from an Excel user merely "getting by," this is it. In Excel, equality rules. Every cell is the same. You type a number, it sits there. You write a formula, it sits there. You can see it, touch it, and audit it. Power BI forces you to abandon this comfort zone. It introduces a divergence that feels counterintuitive, perhaps even wrong at first: **Calculated Columns** versus **Measures**.

They both utilize DAX. They appear identical in the formula bar. But under the hood? They behave like matter and energy. Confusing the two isn't just a syntax error; it is the primary reason your reports lag, your file size bloats, and your numbers turn out mathematically impossible.

18 divided by $2,020.

The Actual Math: 5.8% Margin.

Look at that gap. The report claims nearly **50%** profitability, while the bank account shows less than **6%**. This isn't a rounding error; it is a mathematical hallucination. If you present that 47.5% figure in a board meeting, you aren't just wrong; you are dangerous. This is the **Average of Averages** trap, and Calculated Columns walk you right into it.

The Fix: Calculate on Demand

To get the correct number, you must stop calculating row-by-row and start calculating over the *context*. You need a **Measure**. A Measure does not exist on the row level. It waits for you to define the filter context (the whole company, a specific region, or a single product category) and *then* executes the math on the totals. The DAX Measure looks like this:

```dax
Total Margin % =
DIVIDE(
SUM(Sales_Table[Profit]),
SUM(Sales_Table[Sales])
)
```

Notice the order of operations. We are not averaging the percentages. We are summing the profit first ($118), summing the revenue second ($2,020), and dividing the two resulting aggregates.

Calculated Columns vs. Measures: When to Use Which

If there is a single concept that distinguishes a true Power BI professional from an Excel user merely "getting by," this is it. In Excel, equality rules. Every cell is the same. You type a number, it sits there. You write a formula, it sits there. You can see it, touch it, and audit it. Power BI forces you to abandon this comfort zone.

It introduces a divergence that feels counterintuitive, perhaps even wrong at first: **Calculated Columns** versus **Measures**. They both utilize DAX. They appear identical in the formula bar. But under the hood? They behave like matter and energy. Confusing the two isn't just a syntax error; it is the primary reason your reports lag, your file size bloats, and your numbers turn out mathematically impossible.

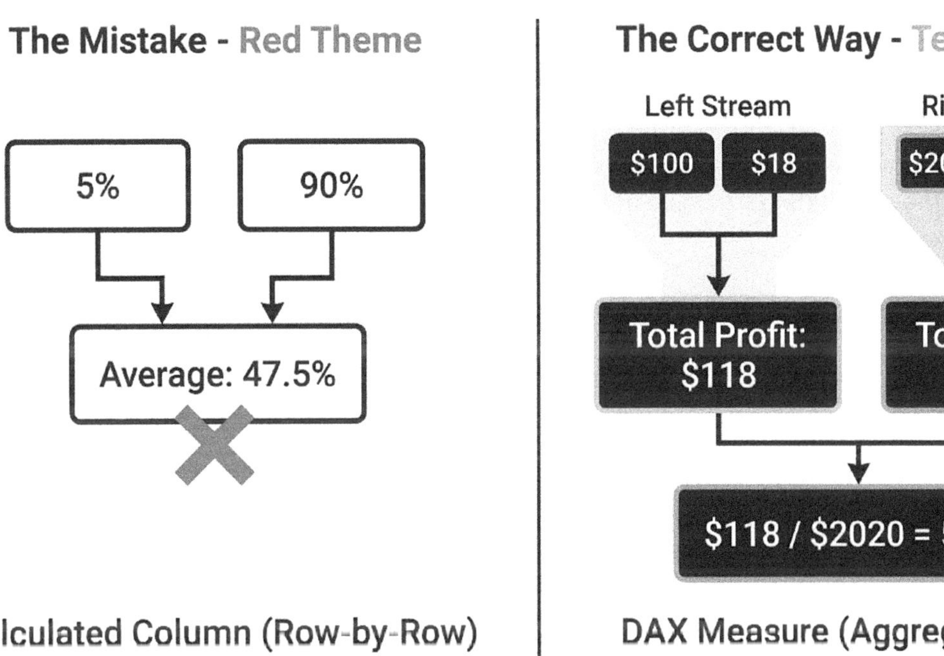

DAY 3

*3.5 - **The Aggregate-First Principle.** A Measure (Right) works differently than a Calculated Column. Instead of doing the math on every row and then averaging the results (which leads to the mathematically false 47.5%), the DAX engine waits. It sums up all the Profit in the filter context first, sums up all the Sales second, and only then performs the division, arriving at the correct 5.8% margin.*

The Golden Rule: If you are calculating a ratio (margin, growth %, density) or a value that cannot be simply added up (like temperature or unit price), **never** use a Calculated Column. Always use a Measure.

The Fundamental Divergence: Static vs. Dynamic

To grasp this, we have to look at *time*. Specifically, when the calculation actually executes.

Calculated Columns are "Hardened Data."

When you write a Calculated Column (e.g., ` Profit = Sales - Cost`), Power BI runs that math immediately. It goes row by row, calculates the result, and stamps it into your RAM and hard drive.

- **The Excel Analogy:** This is exactly like adding a new column in an Excel Table (Column C = A + B) and hitting "Save." The numbers are baked in. They do not change until you refresh the entire dataset.
- **The Mechanism:** Calculated Columns rely on **Row Context**. They have tunnel vision. They look at the specific row they are sitting in, compute the answer, and ignore everything else. They are completely oblivious to the slicers or date filters on your report page.

Measures are "Virtual Logic."

A Measure does not store data. It stores a recipe. When you write a Measure, you aren't baking the cake; you're writing down the instructions. The calculation sits dormant. It does not execute until you drag that Measure onto a visual. Even then, it recalculates every single time a user clicks a slicer.

DAY 3

- **The Excel Analogy:** Think of a Calculated Field in a PivotTable. It doesn't exist in the raw source data; it only manifests in the summary view, reshaping itself based on how you pivot the rows and columns.
- **The Mechanism:** Measures rely on **Filter Context**. They are hyper-aware of their environment—the slicers, the page filters, the axis of the chart—and they calculate an aggregate result for that specific slice of reality.

The Hidden Cost: Storage vs. Speed

Here is the brutal truth about report performance. Beginners rely on Calculated Columns because they can *see* the results in the Data View grid. It feels safe. It feels like a spreadsheet. But this psychological safety comes at a high price.

The Cost of Columns: RAM & Disk

Because Calculated Columns are materialized—meaning saved—data, they bloat your ` .pbix` file. More critically, they consume RAM. Power BI's engine (VertiPaq) is an in-memory database. If you add fifty "helper columns" to a table with 10 million rows, you are forcing your laptop to hold 500 million extra data points in active memory.

- *The Consequence:* Your report takes an eternity to load, and refreshing the data becomes a sluggish nightmare.

The Cost of Measures: CPU

Measures take up zero disk space. They are just text strings. However, they demand processor power at the moment of query.

- *The Consequence:* If your DAX is inefficient and your visual is massive, the user might see the "spinning wheel" while the CPU crunches the numbers. However, the file size remains lean, and the data refresh is instant.

The Decision Matrix: The Golden Rules

To avoid "Technical Debt"—the future headache of untangling a slow, broken model—you need rigorous logic before writing a single line of code. Don't guess. Apply these two tests.

Rule 1: The Axis Test (The "Slice" Rule)

Ask yourself: **"Do I need to put this value on an Axis (X/Y), in a Slicer, or in a Legend?"**

- **If YES:** It **must** be a Calculated Column.
- **The Logic:** You cannot slice data by a Measure. Slicers require a static, finite list of unique values (like "Red", "Blue", "Green" or "High Margin", "Low Margin") to group your rows. A Measure changes constantly; it cannot serve as a fixed anchor for a chart axis.

Rule 2: The Aggregation Test (The "Math" Rule)

Ask yourself: **"Is the result a numerical aggregation (Sum, Average, Ratio, Margin %)?"**

- **If YES:** It **should** be a Measure.
- **The Logic:** Measures calculate correctly at any level of granularity (Grand Total vs. Subtotal). Calculated Columns often fail mathematically when aggregated.

DAY 3

The Truth About Aggregation: Why "Helper Columns" Fail

This is where the "Excel Mindset" betrays the Analyst. In Excel, if you want a profit margin percentage, you create a column `= Profit / Sales` and drag the fill handle down. Then, to get the total, you might accidentally sum or average that column.

Let's look at the **"Ratio Trap."**

Imagine you have two transactions:

1. **Sale A:** Profit \$10, Sales \$100 (Margin = 10%)
2. **Sale B:** Profit \$50, Sales \$100 (Margin = 50%)

The Calculated Column Approach (The Trap):

You create a column `Margin %`. Power BI calculates 10% for row 1 and 50% for row 2. When you drag this column into a visual, Power BI's default behavior is to **SUM** the column.

- *The Result:* 10% + 50% = **60%**.

This is mathematically garbage. You cannot sum percentages. Even if you change the aggregation to "Average," you are averaging the *percentages* (30%), not the weighted totals, which leads to Simpson's Paradox errors.

The Measure Approach (The Solution):

You create a measure: `Margin % = DIVIDE( SUM(Profit), SUM(Sales) )`.

Power BI waits. It sums the total profit (\$60). It sums the total sales (\$200). Then—and only then—does it divide.

- *The Result:* 60 / 200 = **30%**.

This is the correct weighted average.

Summary Heuristic

If you are ever in doubt, memorize this mantra. It will save you from 90% of beginner errors:

"If you Slice by it, Column it.

If you Count it, Measure it."

Resist the urge to see your data in a grid. Trust the virtual nature of the Measure. Your file size will stay small, your math will be accurate, and your report will react instantly to the user's curiosity.

DAY 3

3.3 The Big Three Aggregators: SUM, AVERAGE, and COUNTROWS

You are accustomed to Excel, where calculations are fundamentally spatial. You point to `A1`, add it to `A2`, and dump the result in `A3`. It's tactile. It's visible.

In the Data Model, spatial coordinates do not exist.

Forget "Row 5." Forget "Cell C12." In this environment, there are only columns, tables, and the filters currently squeezing them. Because DAX is blind to specific cells, it cannot simply "look" at a number. It needs instructions on how to process the entire column at once.

This is the role of the **Aggregator**.

Think of an aggregator as a high-powered industrial compactor. You pour thousands—sometimes millions—of rows of raw transactional data into the top, and the function crushes that volume into a single, scalar value that fits neatly into a report card or chart.

We are going to dismantle the foundational triad: **SUM**, **AVERAGE**, and **COUNTROWS**. Their names are identical to the Excel functions you've used for a decade. Do not let that fool you. Their behavior in a database environment requires a complete rewiring of your instincts.

1. SUM: The Unforgiving Accountant

Syntax: `SUM( <Table>[ColumnName] )`

In Excel, the `SUM` function is surprisingly polite. You can accidentally highlight a range containing a mix of numbers, text, and errors, and it often tries to make sense of the mess for you. The DAX engine is not so lenient.

DAX is built for speed over massive datasets. To achieve this, it relies on strict data typing. When you invoke `SUM`, the engine scans the column dictionary expecting pure numeric data—Whole Number, Decimal, or Currency. It does not "guess."

The Friction Point:

Here is where most Excel pros hit a wall: **You cannot SUM a Boolean.**

In Excel modeling, it is standard practice to flag a row as `TRUE` and then sum the column to count how many flags exist (because Excel treats TRUE as 1). In DAX, `SUM(Table[IsActive])` will trigger an immediate error. The engine demands you treat data types with respect. If you try to sum text, the visual breaks.

Imagine `SUM` as a vertical laser beam. It shoots down a single column, slicing through the table, ignoring any rows currently hidden by your filters, and accumulating the value of what remains visible. It creates a "Grand Total" on the fly, every time a user clicks a slicer. Note that `SUM` is chemically inert to `BLANK` values; it treats them as neutral ghosts. `5 + BLANK` is simply `5`.

2. AVERAGE: The Silent Trap

Syntax: `AVERAGE( <Table>[ColumnName] )`

This is the single most common source of "quiet errors" in business intelligence. These are the worst kind of errors—the report doesn't crash, the visual renders perfectly, but the number is wrong.

An average is mathematically simple: `Numerator (Sum) / Denominator (Count)`. The danger lies entirely in the **Denominator**. In a database, there is a profound, non-negotiable difference between "0" and "(Blank)."

- **0 (Zero):** A valid data point representing "no activity." It increases the denominator.
- **BLANK (Null):** The absence of data. It is invisible to the denominator.

The Reality Check:

If a salesperson made zero calls yesterday, but the data was entered as `NULL` (or left empty) instead of `0` in your source system, DAX will **artificially improve** their average performance. It pretends that day never happened.

Excel users often operate on the assumption that "Empty = 0." DAX does not. If you want Blanks to drag the average down, you must physically replace them with zeros in Power Query or use a divisor measure like `DIVIDE`.

3. COUNTROWS: The Structural Shift

Syntax: `COUNTROWS( <Table> )`

This function represents your graduation from "Spreadsheet Thinking" to "Model Thinking."

In Excel, if you want to count how many orders you have, you instinctively reach for a column—perhaps the "Order ID" column—and highlight it to see the count in the status bar. You are conditioned to count cells. In DAX, we stop looking at columns. We look at the **entity** itself. We don't count the "Order IDs"; we count the *Rows* of the "Sales" table.

Why change?

`COUNTROWS` is the single most efficient way to measure volume. It doesn't need to load a specific column into memory to check for values; it simply queries the table's internal statistics to see how many rows exist in the current filter context. It is lighter on the processor (CPU) than counting a specific column. More importantly, it is safer. Stop using `COUNT(<Column>)` unless you have a very specific reason.

Why? Because `COUNT(Sales[OrderDate])` will silently fail you if a legacy system imported a sale with a missing date. The sale happened—the row exists—but your measure won't count it because the specific column you pointed at was blank. `COUNTROWS(Sales)` is fail-safe. It counts the transaction regardless of data quality issues in specific columns.

Function	What you give it	What it returns	The "Gotcha"
SUM	A Column	Total value	Fails on Boolean/Text; ignores Blanks completely.
AVERAGE	A Column	Arithmetic Mean	Ignores Blanks in the denominator, leading to accidental metric inflation.
COUNTROWS	A Table	Integer count	Requires a Table, not a column. The fastest, safest way to count volume.

3.4 Logical Functions: IF, SWITCH, and RELATED

The Logic Engine: Making Decisions with Data

Data without context is just noise. In Excel, you likely live in a world of nested `IF` statements to categorize rows or flag exceptions. In DAX, logical functions serve the same purpose: they act as the decision-making engine of your model, allowing you to segment customers, prioritize orders, and classify products dynamically.

Let's apply this to **SmartGear Retail**. We aren't just calculating totals; we are building business intelligence to distinguish between a high-value laptop sale and a bulk order of USB cables.

IF: Handling Binary Decisions

The `IF` function in DAX is nearly identical to Excel. It checks a condition and returns one result if true, and another if false.

The SmartGear Scenario:

The Logistics Manager needs a quick way to flag orders that require **Priority Shipping**. Policy states that any single transaction exceeding **$500** (like a laptop purchase) qualifies for expedited handling. Standard accessory orders (like cables) go via standard ground shipping. Instead of manually reviewing thousands of rows, we create a Calculated Column in the `Sales` table:

```dax
Shipping Priority =
IF(
Sales[Total Amount] > 500,
"Priority Express",
"Standard Ground"
)
```

DAY 3

This simple logic instantly segments your fulfillment process. However, when business rules get complex—for example, if you have five different shipping tiers based on margin, weight, and region—nested `IF` statements become unreadable 'spaghetti code'. That is where `SWITCH` takes over.

SWITCH: The Elegant Alternative to Nested IFs

Think of `SWITCH` as a cleaner, more readable way to handle multiple conditions. While it can map values 1-to-1 (like changing "1" to "January"), its real power for analysts lies in the `SWITCH(TRUE(), ...)` pattern. This allows you to evaluate different logical tests line by line.

The SmartGear Scenario:

The Finance team wants to classify sales not just by product name, but by **Strategic Role**.

- **High Margin Accessories:** Products with a margin > 40% (e.g., Cables, Chargers). These are your profit generators.
- **Volume Drivers:** Products with a margin < 10% (e.g., Laptops). These drive revenue and acquire customers but have thin margins.
- **Standard Portfolio:** Everything else.

Here is how we translate that strategy into DAX:

```dax
Product Strategy Category =
SWITCH(
TRUE(),
```

```
'Product'[Margin %] > 0.40, "High Margin Accessory",
'Product'[Margin %] < 0.10, "Hardware / Volume Driver",
"Standard Portfolio"
)
```

This approach is self-documenting. A year from now, when you revisit this code, you will immediately understand the business logic without deciphering twenty parentheses.

RELATED: The "VLOOKUP" of the Data Model

In Excel, if you are in a Sales table and need the Product Cost to calculate profit, you use `VLOOKUP` or `XLOOKUP`. In Power BI, tables are separate. The `Sales` table (Fact) knows *how many* items were sold, but it doesn't know the *standard cost* of the item—that lives in the `Product` table (Dimension). Since `Sales` and `Product` are connected by a relationship, DAX uses `RELATED` to traverse that wire.

DAY 3

The SmartGear Scenario:

We need to calculate the **Gross Profit** for every transaction in the `Sales` table.

- *Formula:* (Sales Price - Product Cost) * Quantity.
- *Problem:* The "Product Cost" column is in the other table.

We use `RELATED` to fetch the cost from the 'One' side (Product) down to the 'Many' side (Sales). Think of this as water flowing downhill: data flows effortlessly from the lookup table down to the transactions.

```dax
Total Profit =
Sales[Quantity] * (
Sales[Unit Price] - RELATED('Product'[Standard Cost])
)
```

Key Takeaway: `RELATED` only works when you are in a row context (like a Calculated Column or an iterator) and you are pulling data from the "One" side of a relationship. It is the bridge that turns isolated tables into a cohesive model.

3.5 Organizing and Formatting Your Measures Table

In Excel, calculations have a physical address. You know exactly where your margin percentage lives: cell F12 on the "Summary" tab. If you move it, Excel tracks it. If you delete the row, the model breaks. Power BI is different.

Here, a measure is a **floating formula**. It has no physical row and no physical column. It is pure logic, suspended in the data model. But the interface forces every measure to have a "Home Table" simply for display purposes. It has to live *somewhere*. By default, Power BI acts like a disorganized intern. If you write a `[Total Sales]` measure while your mouse happens to be selecting the `Customers` table, the software dumps the measure there. If you create `[Total Costs]` while clicking on `Date`, that's where it stays.

The Brutal Truth: If you rely on this default behavior, your field list will succumb to "Spaghettification." You will end up hunting for critical financial metrics inside your `Employee` table. As a knowledge worker, your cognitive load should be spent on high-level analysis, not a game of hide-and-seek. We need to borrow a concept from database architecture: **Decoupling Logic from Data.** We are going to build a dedicated VIP lounge for your DAX formulas.

The "Enter Data" Hack

Microsoft does not (yet) provide a native "Create Measure Table" button. We have to engineer one. We use a specific protocol known as the **Empty Table Hack**. This creates a central repository that holds *only* your calculations, keeping your raw data tables pristine.

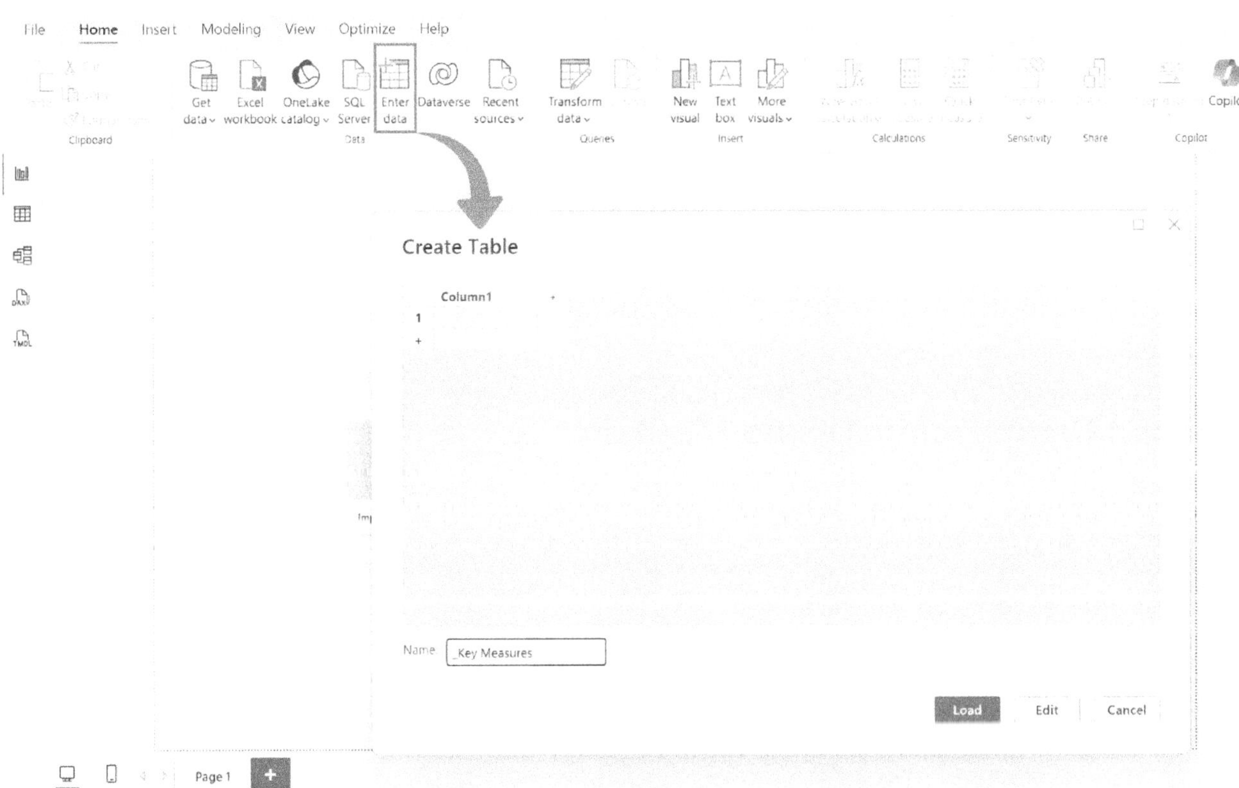

*3.6 - **Creating the Empty Folder.** To build a dedicated home for your DAX, start by clicking Enter Data and creating a blank table. Crucially, name it '_Key Measures'. The underscore is not a typo; it forces Power BI to sort this table alphabetically at the very top of your Fields pane, keeping your logic separated from your raw data.*

The Protocol:

1. **Action:** Go to the `Home` ribbon and click **Enter Data**.

2. **The "Nothing" Step:** A grid appears, looking like a mini-Excel sheet. **Do not type anything.** Leave the grid completely empty.

3. **Naming Strategy:** In the Name field at the bottom, type `_Key Measures`.

    - *Why the underscore?* Power BI sorts tables alphabetically. The underscore (`_`) forces this table to the absolute top of your list. Your metrics should always be the first thing you see.

4. **Load:** Click **Load**.

The Icon Transformation

Right now, the engine is confused. It sees `_Key Measures` as just another table containing data. It assigns it the standard "grid" icon. To tell Power BI, "This is a container for logic, not data," we must purge it of physical columns.

DAY 3

1. **Move a Measure:** Create a test measure or select an existing one (e.g., `Total Sales`). In the `Measure Tools` ribbon, change the **Home Table** dropdown to `_Key Measures`.

2. **The Purge:** Go to the **Data pane** (on the right). You will see your measure sitting next to a generic column named `Column1` (automatically created by Power BI in the previous step).

3. **Delete the Artifact:** Right-click `Column1` and select **Delete from Model.**

4. **The Metamorphosis:** The moment the last physical column is removed and only measures remain, the table icon changes from a **Grid** to a **Calculator Stack.**

This isn't just cosmetic. It is a visual signal that you have successfully separated the *physics* of your model (rows and columns) from the *logic* (measures).

Advanced Organization: Display Folders

In Excel, you manage complexity with tabs. In Power BI, as you scale from ten measures to fifty, a flat list becomes unusable. You need folders.

The Professional's Workflow (Model View):

Don't do this in Report View; it's too slow. The **Model View** is your command center for architectural changes.

1. **Select:** Click on a measure (or multiple measures using `Ctrl`) in the Model View.

2. **Locate:** In the **Properties** pane, find the field labeled **Display Folder.**

3. **Define:** Type the name of the folder, for example, `Financials`.

The "Sub-Folder" Syntax:

You can create hierarchy without clicking buttons. Use a backslash (`\`) to create nested folders instantly.

- Typing `Financials\Ratios` forces the measure into a `Ratios` folder *inside* a `Financials` folder.
- Typing `Sales;Financials` (using a semicolon) puts the *same* measure in two different folders simultaneously—a feat impossible in standard file systems.

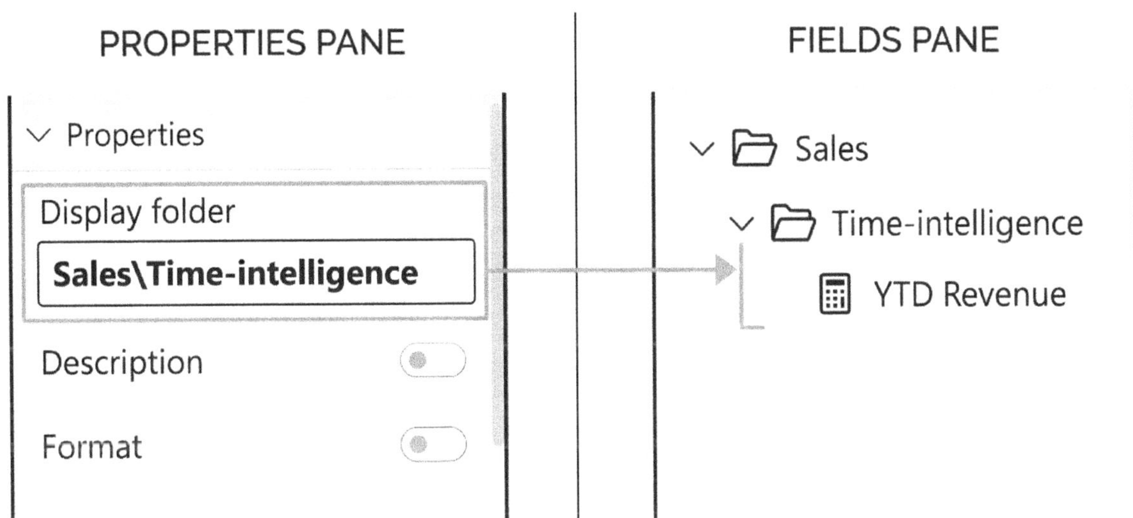

*3.7 - **Creating Sub-Folders.** By typing Sales\Time-intelligence in the Display folder field (Left), you force Power BI to create a folder structure (Right). The backslash '\' acts as a directory separator, allowing you to organize dozens of measures into logical categories, rather than leaving them in a chaotic alphabetical list.*

DAY 3

Bulk Formatting (The Efficiency Hack)

Here is where the analyst in you should revolt against manual labor. Formatting measures one by one—setting Currency, 2 decimals, thousands separators—is a waste of billable hours.

Use the **Model View** for batch processing:

1. **Multi-Select:** Click the first measure, hold `Shift` (for a range) or `Ctrl` (for specific items), and select every measure that represents money.
2. **Apply Once:** In the Properties pane, change **Format** to `Currency` and **Decimal Places** to `0`.
3. **Propagate:** Power BI applies this setting to all 20 selected measures instantly.

A Note on Safety (Excel Trauma vs. DAX Reality)

If you have spent years in Excel, you likely suffer from "Reference Anxiety." You know that dragging a cell from A1 to Z99 can explode your spreadsheet with `#REF!` errors.

Relax. DAX is resilient.

DAX references are semantic (based on names), not geographic (based on location). You can move `[Total Sales]` from the `Orders` table to your new `_Key Measures` table, rename the folder, or shuffle it around endlessly. Your charts, KPIs, and reports will **not** break. The visuals only care *that* the measure exists, not where it sleeps at night.

Day 4 - Advanced DAX: Answering Real Business Questions

4.1 CALCULATE: The Most Powerful Function in Power BI

The Anatomy of Control

If DAX had a monarch, this would be it.

Forget `VLOOKUP`. Forget everything you know about standard aggregation. If you learn nothing else from this entire guide, learn this function. In Excel, `SUMIFS` feels safe. It's logical. You pick a range, you set a criteria, you pick a sum range. Done. But let's be honest: `SUMIFS` is brittle. It snaps the moment you add a new dimension to your analysis. It becomes a maintenance nightmare the second you need complex logic.

CALCULATE is the answer. It isn't just a "super-SUMIFS"; it is the engine that separates basic reporting from actual business intelligence.

DAY 4

The Director on the Set

To really get this, you have to accept a harsh truth about your data: **Every number you see in a Power BI report is biased.**

It's biased by the "Filter Context." That \$1M sales figure for "Red Products" in "2023"? It is being bullied by the slicers on your page and the headers in your matrix. Standard DAX functions like `SUM` or `AVERAGE` are passive actors: they simply read the script given to them by the filters.

CALCULATE is the Director.

It is the only function with the authority to walk onto the set, scream "Cut!", rearrange the props, change the lighting, remove an actor, and *then* tell the camera to roll.

Technically, it is an **environment controller**. It grants you three god-like powers over your data model:

1. **Add** a filter the user didn't select (e.g., "Calculate Sales, assuming the Country is *always* USA").

2. **Remove** a filter the user *did* select (e.g., "Calculate Sales, ignoring whatever Color the user clicked").

3. **Overwrite** an existing filter (e.g., "The user selected 2024, but I want to calculate this specific metric using 2023 data").

The Syntax: Deceptively Simple

The syntax looks innocent enough. Don't be fooled.

```dax
CALCULATE( <Expression>, <Filter1>, <Filter2>, ... )
```

There are two main components here:

1. The Expression (Parameter 1)

This is the calculation you want to run. **Crucial advice:** It should almost always be a pre-existing measure, like `[Total Sales]`.

- *The Pro Tip:* Never write raw formulas like `SUM(Table[Column])` inside a CALCULATE if you can avoid it. Use a Measure. If you change the logic of `[Total Sales]` six months from now, your CALCULATE function updates automatically. We call this **Measure Branching**. It saves lives.

2. The Filters (Parameter 2+)

These are your instructions to alter reality. You can pass simple constraints like `Sales[Country] = "USA"` or use advanced table modifiers like `ALL` or `FILTER`.

The Hidden Algorithm

Here is where Excel veterans usually trip up. Excel formulas generally calculate left-to-right. DAX? It follows a rigid, four-step algorithm to determine the "truth" of your number.

You aren't just writing a formula; you are programming a sequence of events.

1. **Context Transition (The Setup):** If you use CALCULATE inside a calculated column, it first takes the "Row Context" (the specific row you are on) and transforms it into a filter. (We'll cover this trap in a second).

2. **Filter Evaluation:** DAX looks at your filter arguments (e.g., `Year = 2023`) and evaluates them in the *original* environment.

3. **Context Modification (The Override):** This is the magic moment. CALCULATE applies your new filters to the model.

   - **The Overwrite Rule:** This is critical. If your report page is filtered to "Year = 2024", but your CALCULATE function says `Year = 2023`, **CALCULATE wins.** It overwrites the page filter. This is exactly how we do Year-over-Year comparisons without manual pivoting.

4. **Expression Evaluation:** Only now—after the environment has been rigged—does DAX run the initial expression (`[Total Sales]`).

The "Context Transition" Trap

This concept separates the novices from the pros.

The Problem: You have likely tried to create a Calculated Column to generate a ratio, perhaps `Sales / Target`. You write the formula, hit enter, and suddenly every single row shows the exact same number—usually the Grand Total. You check your math. The math is fine. The **context** is wrong.

The Solution: Measures are blind to the row they are currently sitting in. They calculate aggregates over the whole dataset. However, when you invoke **Context Transition**, you force the measure to "see" the row it is on and filter the data to *only* that row.

Rule of Thumb: Whenever you drop a Measure into a Calculated Column formula, DAX automatically wraps it in a hidden CALCULATE to trigger this transition. It bridges the gap between "Row World" (Excel thinking) and "Filter World" (Database thinking).

Real World Application: The "Percentage of Total"

Let's get practical. A classic business question: *What is the market share of our products?*

In Excel, you would create a Pivot Table, calculate the sum, copy-paste it to a side column, and divide. It's manual. It's static. It's prone to error. In Power BI, if you simply divide `[Total Sales]` by `[Total Sales]`, the result is always 100%. Why? Because if you filter to "Red Products," both the numerator and the denominator are filtered to "Red Products."

You need a denominator that **ignores** the visual's filter.


```dax
% of Total =

DIVIDE(

[Total Sales],

CALCULATE( [Total Sales], ALL( Sales ) )

)
```

The Mechanism:

1. **Numerator:** `[Total Sales]` respects the user's selection (e.g., "Red").

2. **Denominator:** `CALCULATE` steps in. The `ALL( Sales )` function strips away all filters from the Sales table. It forces the result to be the Grand Total of all sales, regardless of what color the user clicked.

3. **Result:** You get "Red Sales" divided by "All Sales." Dynamic. Instant.

Real World Application: The Time Machine

The Scenario: Your CFO asks, "How are we doing compared to this time last year?"

The Challenge: Your report is filtered to "January 2024." To answer the question, you need data from "January 2023" *without* asking the user to touch the slicer.

The Solution: Use CALCULATE to shift the time window.

```dax
Sales PY (Previous Year) =
CALCULATE(
[Total Sales],
SAMEPERIODLASTYEAR( 'Date'[Date] )
)
```

How it works:

1. CALCULATE pauses the operation.
2. It looks at the current filter ("January 2024").
3. It passes this date to `SAMEPERIODLASTYEAR`, which returns a list of dates for "January 2023."
4. CALCULATE **overwrites** the filter context. The model now "thinks" the date is January 2023.
5. It runs `[Total Sales]` in this past environment.

DAY 4

This allows you to place "Sales 2024" and "Sales 2023" side-by-side in the same visual. No manual VLOOKUPs. No extra spreadsheets. Just logic.

4.2 The Business Case: Solving the "Year-Over-Year" Problem

The Excel Headache vs. The Power BI Magic

Before we dive deeper into syntax, let's pause. Why should you care about `CALCULATE`? Why struggle with "Context Transition"?

The answer lies in the most common request every analyst receives: **"How did we do compared to last year?"**

In the Excel world, calculating Year-Over-Year (YoY) growth is a manual nightmare. You have to create a pivot table for 2023. Then another column for 2024. Then a third column with a formula `=(2024-2023)/2023`. It works fine until the boss says, "Great, now show me that by Region."

Your formula breaks. The rows shift. You spend the next hour re-aligning cells and fixing `#DIV/0!` errors. It is fragile, static, and prone to human error.

Power BI solves this with Time Intelligence.

Instead of building new columns, we use `CALCULATE` to shift time itself. We create a measure that says: *"Calculate Total Sales, but pretend the date is exactly one year ago."*

The "Same Period Last Year" Pattern

This is the bread and butter of financial reporting. We will use a dedicated function called `SAMEPERIODLASTYEAR`. It works hand-in-hand with `CALCULATE` to perform a "temporal shift."

The Goal:

We want to see "Sales 2024" and "Sales 2023" side-by-side in the same visual, responding dynamically to any slicer.

Step 1: The Base Measure

We already have our core metric:

`Total Sales = SUM(Sales[Amount])`

Step 2: The Time Shift

Now, we create the comparison measure using `CALCULATE`.

4.3 Time Intelligence: YTD, MTD, and Same Period Last Year

You have just mastered the annual comparison using `SAMEPERIODLASTYEAR` in the previous section. That is perfect for high-level annual reviews. However, businesses rarely run on annual autopilot; they are managed day by day and month by month.

To manage the daily grind, you need two new capabilities: the ability to track cumulative progress (Year-to-Date) and the flexibility to compare against *any* previous timeframe (like Month-over-Month), not just last year.

Before we hand you these new tools, we must address the infrastructure they run on. Without this specific foundation, the functions below will fail

The Non-Negotiable: A Continuous Date Table

This is the single most common point of failure for beginners moving from Excel to Power BI. **Time Intelligence functions do not work reliably on your Sales table.**

In Excel, if you don't have sales on January 1st, you simply don't type a row for January 1st. Excel doesn't care. DAX, however, treats time as a mathematical continuum. When it tries to calculate a year-to-date total or shift back one month, it requires a complete timeline to iterate over.

If you run these functions directly on your `Sales[OrderDate]`, you are using a "Swiss Cheese" calendar—full of holes where no transactions occurred. The DAX engine will fall into these gaps, returning errors or, worse, subtly incorrect numbers that go unnoticed until a finance meeting.

The Rules of Engagement:

1. **Separate Table:** Do not use the date column in your sales data. You must connect a dedicated `'Date'` table to your model.

2. **No Gaps:** This table must have a row for *every single day* of the year (Jan 1 to Dec 31), weekends and holidays included, even if your shop was closed.

3. **Full Years:** It must cover the full fiscal year. If your sales data starts in March, your Date table must still begin on January 1st.

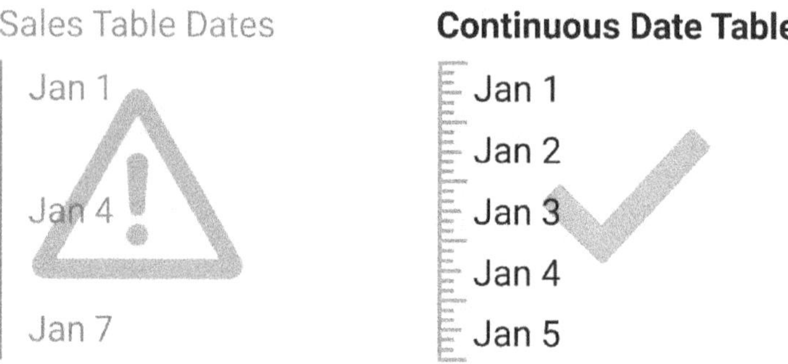

*4.1 - **The Gap Trap.** If your store closes on January 2nd and 3rd, those dates simply don't exist in your Sales Table (Left). If you ask Power BI to calculate "Previous Day" sales using that table, it will break. A dedicated Date Table (Right) acts as an unbroken timeline, allowing Time Intelligence functions to travel backward safely, even over weekends and holidays.*

DAY 4

The "Trip Odometer": TOTALYTD

The most frequent request in reporting is the running total: *"Where do we stand right now?"*

In Excel, creating a Year-to-Date (YTD) view requires fragile formulas that you must manually reset when the new year starts. Power BI handles this with `TOTALYTD`.

Think of this function as your car's **trip odometer**. As you drive through the year (Jan 1 to Dec 31), the mileage counts up cumulatively. The split second the calendar flips to the new year, the engine automatically snaps the odometer back to zero and starts counting again.

The Syntax:

Power BI offers us "Syntactic Sugar"—a simplified shortcut for this complex logic:

```dax
Sales YTD = TOTALYTD( [Total Sales], 'Date'[Date] )
```

How It Works:

1. **Scan:** It looks at the date filter on your report (e.g., "March 15, 2024").

2. **Reset:** It automatically finds the start of that year (January 1, 2024).

3. **Sum:** It adds up every transaction between the start line and your current date.

Note for Controllers: `TOTALYTD` assumes a standard Gregorian calendar. If your company operates on a non-standard Fiscal Year (e.g., starting July 1st), you must add the optional year-end parameter: `TOTALYTD(..., "06-30")`.

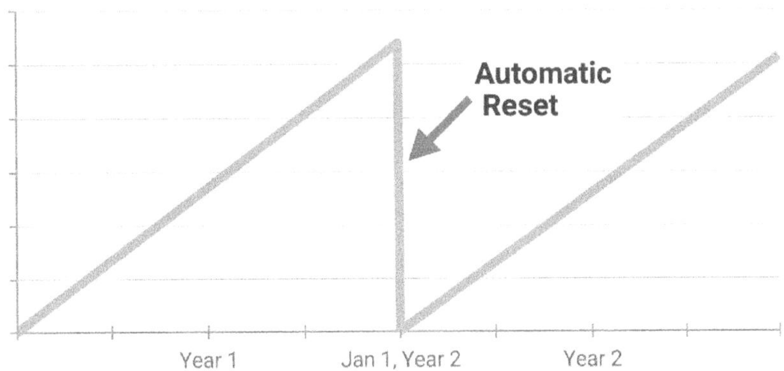

*4.2 - **The Cumulative Context.** TOTALYTD is a specialized form of CALCULATE. It automatically modifies the filter context, forcing the engine to sum not just the current month, but all months since the beginning of the calendar year, resetting annually.*

The Upgrade: DATEADD vs. SAMEPERIODLASTYEAR

In the previous section, we used `SAMEPERIODLASTYEAR`. While useful, it is a "one-trick pony." It can only travel back exactly one year. What if your VP asks: *"How are we trending compared to last month?"* For this, you need the universal time machine: **DATEADD**.

DAY 4

`DATEADD` allows you to shift your current context backward or forward by any interval you choose (Days, Months, Quarters, or Years). It is the standard tool for Month-over-Month (MoM) analysis.

```dax
Sales Last Month =
CALCULATE(
[Total Sales],
DATEADD( 'Date'[Date], -1, MONTH )
)
```

The "Window Shift" Concept:

Crucially, `DATEADD` is smart about window sizing. It doesn't just subtract 30 days; it respects the *duration* of your selection.

- If you select the **entire month of June,** `DATEADD` shifts back to the **entire month of May.**
- If you select just **June 1st to June 10th,** `DATEADD` shifts the window to **May 1st to May 10th.**

It preserves the size of the window but moves its position in time.

The Danger Zone: The Partial Period Trap

There is a classic beginner mistake that destroys credibility in monthly reporting. We call it the **Partial Period Trap**.

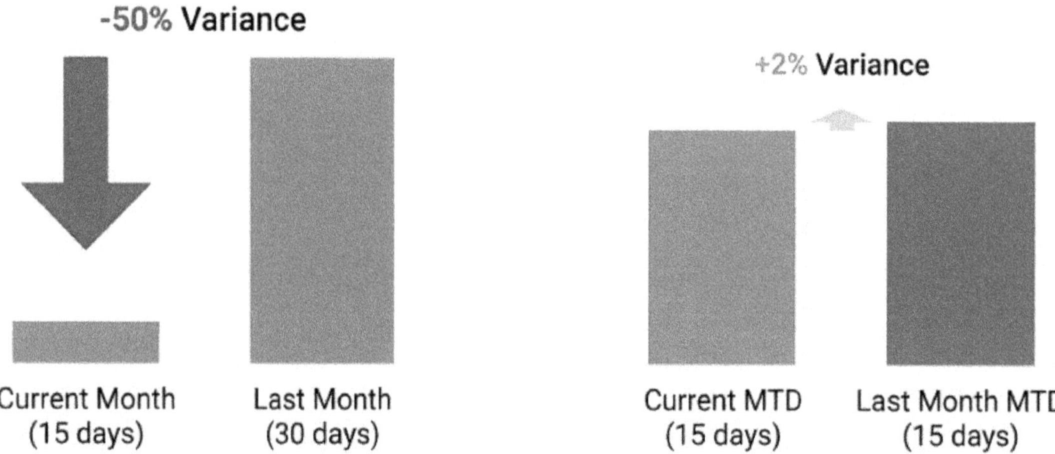

*4.3 - **The Partial Period Trap.** If you check your dashboard on the 15th of the month, comparing it to the entirety of last month will always show a massive, terrifying drop in sales (Left). Time Intelligence functions like TOTALMTD automatically slice last month's data to match the current day, ensuring an "apples-to-apples" comparison (Right).*

DAY 4

The Scenario:

It is **October 15th**. You refresh your report to check progress. `[Total Sales]` calculates revenue for the first 15 days of October. If you compare this blindly against `DATEADD(..., -1, MONTH)`, DAX looks at the "Previous Month" context. Since the current month is October, the previous month is September. DAX compares your **15 days** of October sales against **30 days** of September sales.

The Result:

Your dashboard shows a massive, terrifying drop in performance (e.g., -50%). The data isn't technically wrong, but the comparison is **factually unfair**. You are comparing a half-finished race to a finished one.

The Solution:

Be aware of this when designing dashboards. When viewing the current incomplete month, you cannot use standard whole-month shifts. You must ensure you are comparing "Month-to-Date" (MTD) vs "Last Month-to-Date" (LMTD) to keep the comparison apples-to-apples.

4.4 Calculating Variances and Growth Percentages

Step 1: The "Absolute" Variance ($)

We know where we are (`[Total Sales]`) and we know where we were (`[Sales PY]`). However, raw totals are rarely enough. A ten-million-dollar month is meaningless without context. If the target was five million, you are a hero. If the target was twenty, you have a crisis. The most critical question in any business review is simple: **"Are we winning?"** To answer this, we need the **magnitude** of the change. In financial modeling, we call this the Variance, or the "Delta."

The Logic

The math is elementary subtraction: `Current State - Previous State`.

The DAX Pattern

```dax
Sales Variance = [Total Sales] - [Sales PY]
```

The "Blank" Advantage

Here is where your Excel muscle memory might make you nervous. In a spreadsheet, if you subtract a number from an empty cell, the result is often unpredictable or requires messy cleanup. You might get a `#VALUE!` error or have to write logic to treat blanks as zeros. DAX is smarter. It handles `BLANK` with elegance. Consider a new product launch. Last year's sales (`[Sales PY]`) do not exist—they are `BLANK`. The calculation effectively becomes `[Total Sales] - 0`. The result is simply `[Total Sales]`. This represents "Pure Growth." You do not need complex `IF` statements to handle new business scenarios. The engine handles the math for you.

Step 2: The "Relative" Variance (% Growth)

This is the section where traditional Excel habits can sabotage your report.

In the spreadsheet world, calculating growth is a minefield. You write a formula like `=(C2-B2)/B2`. You drag it down ten thousand rows. Suddenly, your pristine report is riddled with `#DIV/0!` errors because a handful of products had zero sales last year. You then waste valuable time wrapping everything in `IFERROR(..., 0)`. Stop cleaning errors manually. Let the engine do it.

The "Airbag" Function: DIVIDE

Power BI offers a specialized function to handle this specific fragility: **`DIVIDE`**. Do not use the forward slash (`/`) for ratios in measures. The forward slash is mathematically ruthless; if the denominator is zero, it breaks the visualization or returns `Infinity`. The `DIVIDE` function is your safety net. It performs the division, checks the denominator, and handles the exception—all in one optimized step.

The Syntax

`DIVIDE( <Numerator>, <Denominator>, [AlternateResult] )`

The DAX Pattern

```dax
```

```
Sales Growth % = DIVIDE( [Sales Variance], [Sales PY], 0 )
```

Why this is superior

1. **The Safety Valve:** If `[Sales PY]` is 0, the function automatically returns the third argument (`0`). This ensures your dashboard shows "0%" instead of crashing.

2. **Engine Optimization:** The DAX engine executes `DIVIDE` significantly faster than a manual `IF( [Sales PY] = 0... )` structure.

Step 3: Professional Polish (Custom Formatting)

Calculating the number is only half the job. As an analyst, your goal is to reduce **Cognitive Load**. When a stakeholder scans a dashboard, their brain looks for patterns before reading digits. A list of numbers like `5%`, `-3%`, `2%` is difficult to process quickly. Compare that to `+5%`, `-3%`, `+2%`. The explicit plus sign acts as a visual anchor, instantly categorizing the value as "Good" before the user even parses the number. By default, Power BI only shows the sign for negative numbers. We need to force the positive sign.

The Trap to Avoid (Read this carefully)

Do **NOT** use the `FORMAT()` function inside your DAX measure (e.g., `FORMAT([Measure], "+0%")`).

DAY 4

If you do this, your measure converts from a **Number** to **Text**.

- **The Consequence:** You can no longer use that measure in charts (bar charts cannot plot text). You cannot sort by it numerically (it will sort alphabetically, where "+10%" comes before "+2%").
- **The Fix:** Change the **Model Formatting**, not the Data Type.

How to apply Custom Formatting strings:

1. Select your `[Sales Growth %]` measure in the Fields pane.

2. Navigate to the **Measure Tools** tab in the ribbon.

3. Locate the "Format" dropdown and select **Custom** (you may need to type it in directly).

4. In the format box, type this exact pattern:

`+0.0%;-0.0%;0.0%`

Decoding the Syntax

This uses a three-part syntax separated by semicolons:

1. **Positive Rule (`+0.0%`):** We force a `+` sign here.

2. **Negative Rule (`-0.0%`):** We ensure the negative sign remains.

3. **Zero Rule (`0.0%`):** How to handle a perfect zero.

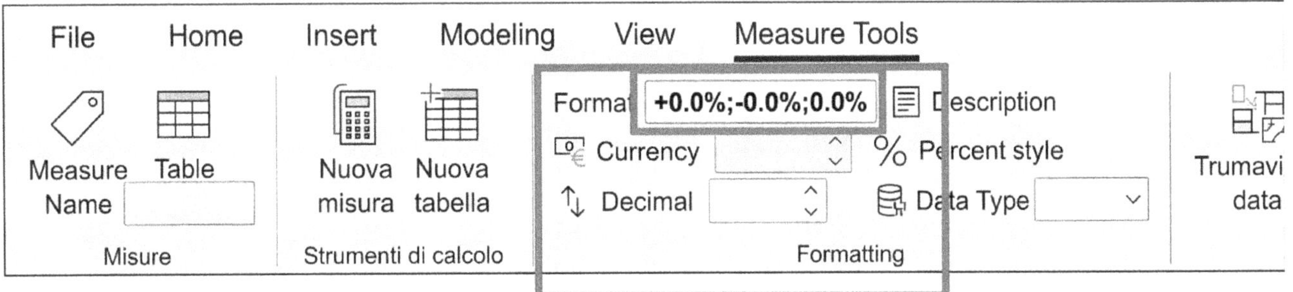

*4.4 - **The Custom Format Hack.** Power BI hides the + sign on positive numbers by default. By typing +0.0%;-0.0%;0.0% directly into the Format bar, you override the default. The syntax has three sections separated by semicolons: how to display Positive numbers, Negative numbers, and Zeroes. This tiny detail drastically reduces the cognitive load for whoever is reading your dashboard.*

The Architect's View

By using this method, the underlying data remains a pure decimal number (0.05). It flows into bar charts, scatter plots, and conditional formatting rules perfectly. The `+` sign is merely a "mask" applied at the very last second when the number is rendered on the screen. This gives you the best of both worlds: the clarity of text and the utility of math.

4.5 Using DIVIDE for Safe Ratios and Margins

The Visual Hazard: How "Infinity" Flattens Your Data

In Excel, a `#DIV/0!` error is annoying—it stains a single cell. In Power BI, that same error is destructive—it breaks the geometry of your report. Here is the mechanics of the disaster: When you divide a number by zero (e.g., `Sales / Units` where Units are 0), the result is mathematically **Infinity**. If this happens inside a Bar Chart, the rendering engine faces an impossible task.

It must scale the Y-Axis to accommodate a value that has no limit. To fit "Infinity" on the screen, Power BI is forced to compress your actual data—millions of dollars in revenue—down to a scale relative to infinity. The result? Your meaningful data collapses into a single, pixel-flat line at the bottom of the chart. One bad calculation destroys the readability of the entire visual.

Strategic Control: The "Ghost" vs. "Zero" Decision

The `DIVIDE` function is your safeguard. It handles the division and, crucially, allows you to dictate what happens when the math fails. This is not just error handling; it is visibility control.

Syntax:

`DIVIDE( <Numerator>, <Denominator>, [AlternateResult] )`

The power lies in the optional third argument: `[AlternateResult]`.

Strategy A: The "Ghost" Strategy (Default)

If you omit the third argument, `DIVIDE` returns `BLANK` when an error occurs.

- **Code:** `DIVIDE( [Total Sales], [Total Units] )`
- **Behavior:** In Power BI visuals, a `BLANK` value effectively does not exist. If a product has no sales, the bar is not rendered. The axis label disappears. The chart auto-filters itself.
- **Best For: Charts and Graphs.** It keeps visuals clean and performant by removing noise.

Strategy B: The "Financial" Strategy (Explicit Zero)

If you provide `0` as the third argument, `DIVIDE` forces a numerical result.

- **Code:** `DIVIDE( [Total Sales], [Total Units], 0 )`
- **Behavior:** The engine returns `0.0`. This forces the data point to exist. In a matrix, you will see a "0" instead of an empty hole.
- **Best For: Financial Tables (P&L).** Controllers often view blank cells as missing data or ETL failures. A hard zero confirms the calculation ran successfully.

The Performance Trap: Why `IF` is Obsolete

DAY 4

Veterans of Excel often try to manually reconstruct error handling using logic like this:

`IF( [Units] = 0, 0, [Sales] / [Units] )`

Stop doing this.

While logical, this approach forces the DAX engine to work twice as hard. In many scenarios, the engine must evaluate the denominator (`[Units]`) once to check if it is zero, and then—if it's not—evaluate it a second time to perform the division. `DIVIDE` is an optimized function built into the core engine (C++). It performs the check and the calculation in a single, highly efficient pass. It is safer, cleaner to read, and faster to execute.

4.6 Handling 'All' and Filter Removal Techniques

The Denominator Problem: Escaping the Filter Context

Think back to your safety blanket in Excel: the dollar sign. When you calculate a "Percentage of Total" in a spreadsheet, you perform a spatial exercise. You have a numerator in cell `B2` and a denominator in `B100`. You lock that denominator down with `$B$100`. You are explicitly commanding the software: "I don't care where I drag this formula; that cell is anchored. Do not move."

In Power BI, this logic collapses instantly. Why? **Because cells do not exist.** There is only Context. This creates a fundamental conflict for anyone migrating from spreadsheets to data models. When you drop a measure like `[Total Sales]` into a matrix, DAX relentlessly filters that measure for every single coordinate in your visual.

If you are looking at the row for "Gaming Laptops," the engine filters the data to that specific subcategory *before* it ever attempts the math. This leads to the classic rookie error. You try a naive formula: `[Total Sales] / [Total Sales]`. The result? **100%. Everywhere.** It is maddening, but logically consistent. DAX filters the numerator to "Gaming Laptops." It also filters the denominator to "Gaming Laptops." You are dividing a number by itself. To solve this, you must manipulate the environment. You must force the denominator to **ignore** the filters that are currently active. You have to tell DAX to look outside the current row.

1. The Bulldozer: ALL()

The `ALL` function is your brute-force method. It is the DAX equivalent of a hard factory reset for your filters.

The Logic:

Mathematically, a ratio requires two different scopes: a local scope (the numerator) and a global scope (the denominator). DAX defaults to local. `ALL` forces the engine to ignore the current filter context entirely and look at the full physical table stored in memory.

The Experience:

Imagine you are in a dark warehouse using a flashlight. The flashlight is your filter. You point the beam at a specific shelf, illuminating only "Audio Equipment." That works for your numerator. But to see the denominator (the total warehouse inventory), you cannot use that focused beam. You need to flip on the overhead floodlights. `ALL` is that floodlight. It reveals everything, regardless of where your flashlight is pointing.

DAY 4

The Critique:

But here is the catch. `ALL` is nuclear. It doesn't just ignore the row filters in your visual; it ignores **everything**. If you have a slicer on the page set to "Year: 2024" and you use `ALL('Sales')`, your denominator will not just be 2024 sales. It will include 2021, 2022, and 2023. It ignores the slicer entirely. This is where beginners panic. They calculate a "Market Share," filter the report to "Europe," and suddenly the percentages drop to 0.5%. You aren't dividing "Europe Sales" by "Europe Total." You are dividing "Europe Sales" by "Global Sales Since the Dawn of Time." `ALL` implies absolute removal of constraints.

Syntax:

```dax
Grand Total Sales =
CALCULATE(
[Total Sales],
ALL( 'Sales' )
)
```

```

| Product Category | Total Sales | Grand Total (ALL) | % of Global |
|---|---|---|---|
| Audio | $10,000 | $100,000 | 10.0% |
| Computers | $50,000 | $100,000 | 50.0% |
| Cameras | $40,000 | $100,000 | 40.0% |
| **Grand Total** | **$100,000** | **$100,000** | **100.0%** |

*4.5 – **The ALL Function in Action.** Look at the middle column. While standard DAX respects the row context (Audio, Computers), wrapping the calculation in ALL(Sales) acts like a bulldozer. It clears the row filters, forcing the engine to return the $100k Grand Total on every single line. This provides the static denominator needed to calculate your % of Global market share.*

## 2. The Intelligent Filter: ALLSELECTED()

Since `ALL` is often too aggressive, we need a function that respects user intent while still ignoring the mechanical row filters of a chart. Enter `ALLSELECTED`.

DAY 4

**The Logic:**

In business analysis, "Total" is rarely "The Total of Everything Ever Recorded." It is usually "The Total of What I Am Currently Looking At." If a Regional Manager at SmartGear filters a report to "North America," they want the denominator to be the "North America Total." They expect the percentages to re-distribute to 100% based on their selection.

**The Experience:**

Think of `ALLSELECTED` as a "Smart Frame." It looks at the report canvas and says: "I will respect the boundaries you set with your external slicers (like Year or Country), but inside this specific visual, I will ignore the breakdown." It allows the user to define the sandbox, but within that sandbox, it grabs the whole pile of sand.

**The Critique:**

`ALLSELECTED` is the function that saves you from angry emails. Nothing confuses a stakeholder more than filtering to a specific region and seeing the "Market Share" column show tiny, irrelevant percentages because the denominator remained fixed on the global total. Users assume that if they filter to "2024," the "100%" should represent 2024 context. `ALLSELECTED` aligns the math with human intuition.

**Syntax:**

```dax
Visual Total Sales =
CALCULATE(
```
```

```
[Total Sales],
ALLSELECTED( 'Sales' )
)
```

3. The Scalpel: ALLEXCEPT()

Sometimes you need surgical precision. You do not want to remove *all* filters, nor do you want to rely solely on what is visible. You need to lock one specific hierarchy while ignoring others.

The Logic:

A common business requirement at SmartGear is "Contribution to Parent." For example, if you are analyzing **Gaming Laptops**, you don't care about their share of Global Sales. You want to know: "What percentage of the **Computers** category do Gaming Laptops represent?". To do this, you must remove the filter on the specific product (Gaming Laptops) while keeping the filter on the wider category (Computers).

The Experience:

Imagine a corporate hierarchy. `ALLEXCEPT` allows you to say, "Forget which specific employee I am looking at, but remember which Department they belong to." It anchors the calculation to the parent node, allowing you to compare siblings against each other.

Syntax:

```dax
Category Total =
CALCULATE(
[Total Sales],
ALLEXCEPT( 'Product', 'Product'[Category] )
)
```

Translation: "Clear all filters on the Product table, **EXCEPT** the filter that defines the Category."

Performance Warning: The "Wide Scan" Risk

I see this mistake in 90% of beginner DAX models. You write `ALL( 'Sales' )`. Technically, this works. Economically, it is expensive. When you use `ALL( 'Sales' )` on a large fact table, you are forcing the engine to scan the *entire* table, every single column, and materialize all unique combinations of rows in memory just to remove the filters. If your SmartGear Sales table has 50 million rows and 40 columns, you are asking DAX to do heavy lifting just to clear a filter on "Color."

The Optimization:

Be specific. If you only need to calculate the percentage of colors, don't unlock the whole table. Unlock only the column you need.

- **Bad (Lazy):** `ALL( 'Product' )` — Scans the whole table.
- **Good (Efficient):** `ALL( 'Product'[Color] )` — Scans only the unique list of colors (perhaps 10 values).

This is the difference between reading an entire encyclopedia to find a definition versus just checking the index. On small datasets, you won't feel it. On real enterprise data, this is the difference between a report that loads instantly and one that spins for 10 seconds.

Day 5 - Visualization: From Data to Insights

5.1 Visual Selection: Matching the Chart to the Message

You have cleaned your data. You have built a robust Star Schema. You have written DAX measures complex enough to replace entire Excel workbooks. Now, you must place pixels on the screen.

Here is the trap most analysts fall into.

They treat visualization as the "decoration phase"—a final coat of paint to make the numbers look pretty. **This is a fundamental error.** Visualization is not art; it is engineering. It is a high-bandwidth mechanism for information transfer. Your sole objective is to move an insight from your computer screen into your stakeholder's brain with the absolute minimum amount of friction.

If your CFO has to squint, guess, or perform mental math to understand a trend, you have failed.

The Biological Hardware: Cognitive Load

To build effective dashboards, you must respect the limitations of your audience. The human brain operates with a limited amount of working memory, or "RAM." We call this **Cognitive Load**.

Every time you ask a user to process a heavy gridline, a decorative border, a 3D shadow, or a redundant legend, you are consuming that RAM. You are burning their mental fuel. If you exhaust their working memory on deciphering the *format*, they have zero energy left to understand the *insight*.

DAY 5

This leads us to the "Data-Ink Ratio," a ruthless metric introduced by Edward Tufte. It poses a simple question: What percentage of the ink (or pixels) on the page is actually representing data?

$$ \text{Data-Ink Ratio} = \frac{ ext{Data-Ink}}{\text{Total Ink Used}} $$

The Brutal Truth:

The default charts you have been using in Excel for the last decade? They usually have a terrible Data-Ink Ratio.

- **Gridlines?** Noise. Delete them or make them a barely visible light grey.
- **Borders?** Clutter. White space is a more elegant separator than a black line.
- **3D Effects?** A lie. A 3D bar chart forces the eye to perform geometry to guess which part of the "block" represents the value. It creates "chart junk" that looks sophisticated to a novice but acts as an obstacle to a professional.

Think of yourself as a sculptor. You do not build a chart by adding elements; you reveal the insight by chiseling away the non-essential.

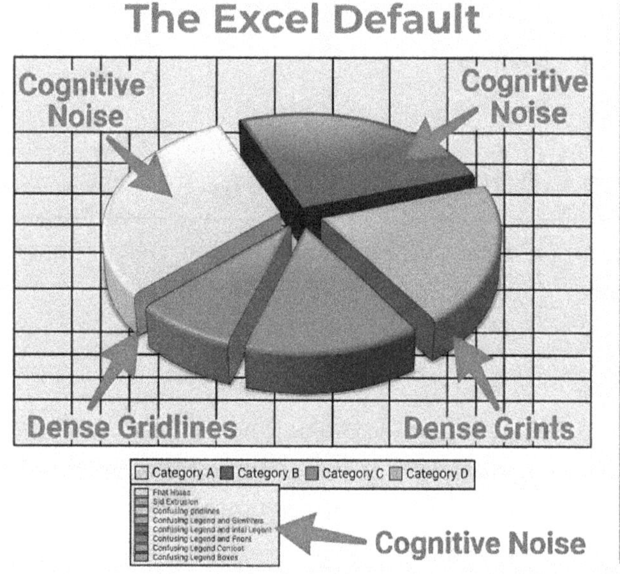

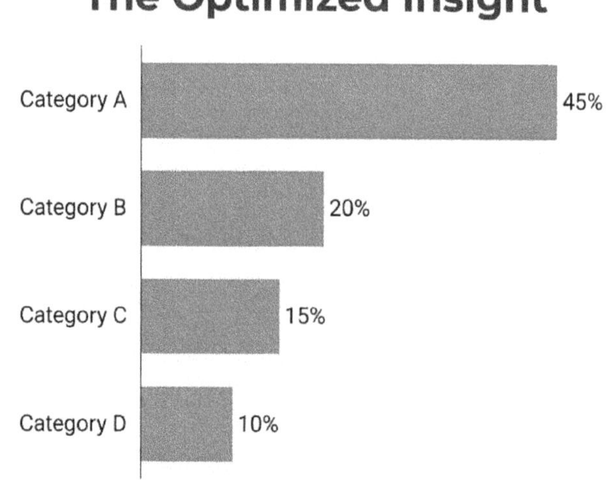

5.1 - ***Eradicating Cognitive Noise.*** *The 3D pie chart (Left) forces the user's brain to constantly dart back and forth between the slices and the legend, while the 3D tilt distorts the actual data proportions. The horizontal bar chart (Right) is the industry standard for a reason. By stripping away the gridlines, the background, and the legend, we maximize the "Data-Ink Ratio." The message is instant and unambiguous.*

The Decision Framework: Intent-Based Selection

Stop asking, "Which chart looks cool?" That is the wrong question. Start asking, "What is the analytical question?"

In a retail environment like SmartGear, 90% of all analytical queries fall into four specific buckets: **Comparison, Composition, Relationship, or Distribution**. We use an adapted version of the Abela Chart Chooser to map these intents to Power BI visuals.

A. Comparison (Ranking and Trends)

This is the "bread and butter" of business analysis. Who sold the most? Which region is underperforming?

- **Scenario 1: Comparing Categories (e.g., Sales by Product)**
- **The Go-To Visual: Clustered Bar Chart (Horizontal).**
- **The Logic:** Why horizontal? Because English is read left-to-right. Consider a typical SmartGear SKU: **"SmartGear Pro-Gaming Laptop 17-inch Gen2"**. If you use vertical columns, you have to tilt the text diagonally or truncate it. You are forcing your user to tilt their head to read the axis. That is friction. Horizontal bars allow long labels to sit naturally. Furthermore, the human eye is exceptionally good at comparing lengths aligned to a common baseline.
- **Scenario 2: Trends Over Time (e.g., Revenue by Month)**

- **The Go-To Visual: Line Chart** (high density) or **Column Chart** (low density).
- **The Rule:** Time flows from left to right. Never, ever put time on the Y-axis. If you have fewer than 12 periods, columns work well to show "weight." If you have 5 years of daily data, a Line Chart is mandatory to show the "shape" of the trend without the visual noise of 1,800 separate bars.

B. Composition (Part-to-Whole)

This is where you show how a total is split. It is also where the most crimes against data are committed.

Let's address the elephant in the room: **The Pie Chart.**

Here is the reality: **Humans are terrible at judging angles.** We struggle to distinguish between a slice representing 25% and one representing 30% without explicit labels.

- **The Rule:** Use a **Donut Chart** only if you have 2 or 3 binary categories. For example, **"Delivery Status (On Time vs. Delayed)"** is acceptable. The distinction is clear and simple.
- **The Alternative:** If you have more than 3 categories (e.g., Sales by Product Category), use a **Treemap** or a **Bar Chart**. A Bar Chart allows for precise comparison of the "parts" much better than a circle ever will.

Composition Over Time:

Use a **Stacked Column Chart**, but be warned. This visual has a flaw. You can easily compare the bottom segment (resting on the x-axis) and the total height. The middle segments? They are floating and jagged. They are notoriously hard to compare across bars. If the relative percentage matters more than the raw volume, switch to a **100% Stacked Chart**.

C. Relationship & Distribution

You are hunting for correlations or outliers in your data.

- **Scenario: Correlation (e.g., Marketing Spend vs. Revenue)**
- **The Go-To Visual: Scatter Plot**. This is the only chart that effectively visualizes two numerical values simultaneously without time being a factor. It instantly reveals clusters and outliers that a table would hide. It can tell you if high ad spend is actually driving high hardware sales, or if you are just burning cash.
- **Scenario: Distribution (e.g., Order count by Basket Size)**
- **The Go-To Visual: Histogram**. In Power BI, this is just a Column Chart, but the "Group" (X-axis) consists of bins (e.g., $0-50, $51-100, $101-150). This shows the "shape" of your transactions—is it a Bell Curve, or heavily skewed toward small accessories?

Power BI Specific "Super-Visuals"

As an Excel user, you are conditioned to think in grids and standard charts. Power BI offers "Super-Visuals" that solve specific dashboarding problems.

- **The Card Visual (The "Headline"):**

In Excel, you bold the cell at the bottom of the table. In Power BI, you extract that number into a **Card Visual**. These are your headlines. Place your most critical KPIs (Revenue, Margin %) at the top-left of the page. This leverages the **"Z-Pattern"**—the natural path the western eye takes when scanning a page.

- **The Matrix (The "Pivot Table on Steroids"):**

Do not use the standard "Table" visual for analysis; it is a flat list. Use the **Matrix**. It allows for hierarchy (drill-downs), stepping layouts, and most importantly, it supports Heat Maps (conditional formatting on the background color). This turns a wall of text into a visual pattern recognition tool.

- **Small Multiples:**

You want to compare sales trends across 12 different countries. In Excel, you create one chart with 12 colored lines. The result is a **"Spaghetti Chart"**—a tangled mess where nothing is readable.

The Solution: Use the **Small Multiples** well in Power BI charts. This takes your one line chart and splits it into a 3x4 grid of clean, individual charts, all sharing the same scale. It is the single best feature for comparing multiple series without clutter.

Critical Rules for Professional Reporting (IBCS Standards)

To be trusted as a Data Analyst, you must adhere to ethical standards. We follow the *International Business Communication Standards* (IBCS) principles. This isn't just about style; it's about integrity.

1. The "Lie Factor" (Truncated Y-Axis)

Do not start your numerical axis at 50 to make a growth from 51 to 55 look massive.

This is common in sensationalist media, but in data analytics, **it is a lie.** If the bar implies volume, the axis *must* start at zero. If you zoom in, you distort the visual proportion, making a 5% increase look like a 200% increase.

Exception: Line charts (trends) can sometimes be truncated if you are clearly focusing on *variance* rather than volume, but proceed with extreme caution.

2. Sort by Magnitude, Not Alphabet

Unless your axis is "Months" or "Days of the Week," never sort alphabetically.

Nobody cares about "Accessories" appearing first just because it starts with "A." They care about which category generates the highest Margin. Always sort categorical charts by the **Value** (Desc or Asc). This creates a "Pareto" shape that instantly highlights your top performers and the long tail.

3. Title Consistency

Don't get creative with titles. Use the format: **"Measure by Dimension"** (e.g., "Total Revenue by Country" or "YTD Margin by Product Category"). This tells the user exactly what they are looking at before they even parse the visual.

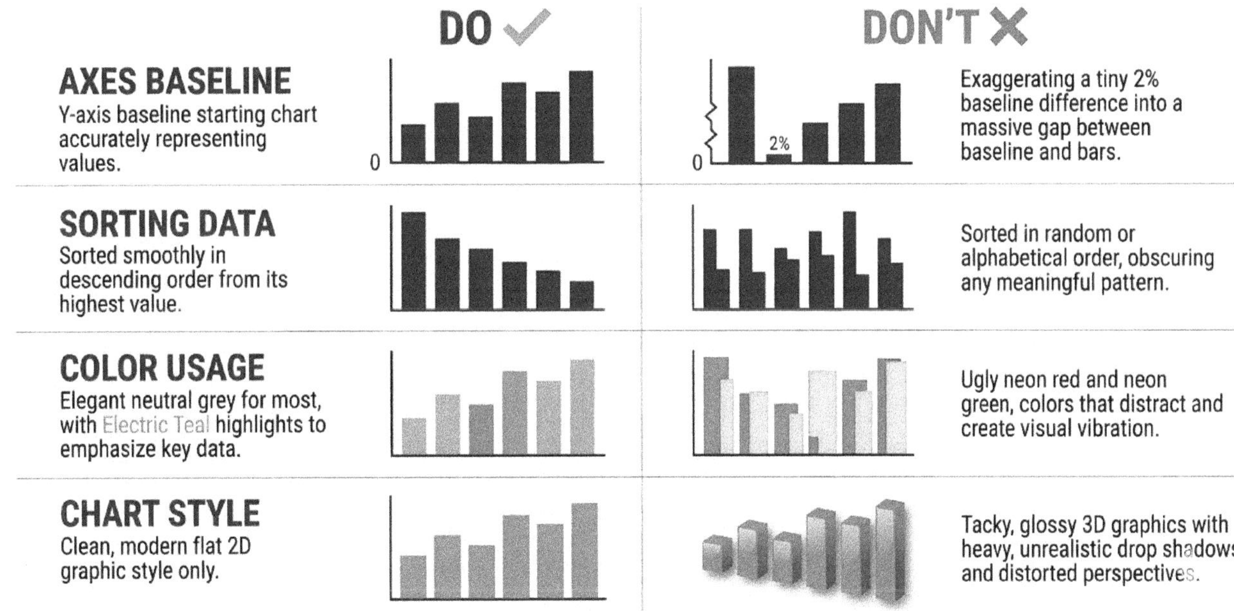

*5.2 - **The Laws of Visual Integrity.** Dashboard design is not about making things "pretty"; it is about protecting the truth of the data. Truncating a Y-axis (Row 1) or using 3D effects (Row 4) physically distorts the numbers, lying to your audience. Random alphabetical sorting (Row 2) forces the brain to work harder to find the winner. Memorize this checklist before you publish your report.*

5.2 The Art of Slicers and Filters: Empowering the End User

In the world of Excel, filtering is a mechanical chore. You hunt for a header, click a tiny arrow, uncheck a box, and watch rows vanish. It is a manual means to an end. In Power BI, however, filtering is not just a utility; it is the **User Interface**. It is the primary mechanism through which your stakeholder asks questions of the data. If that interface feels clunky or fragile, the user assumes the data underneath is broken.

This section shifts focus from the mechanics of hiding rows to the psychology of **User Experience (UX)**. We are building a system that allows a Store Manager at SmartGear to slice through millions of transaction records without friction, transforming a static report into an interactive conversation.

1. The Strategic Distinction: Slicers vs. Filters

Technically, a Slicer and a Filter send the exact same SQL instruction to the database: `WHERE Region = 'North'`. To a human, however, they represent different cognitive modes. One invites exploration; the other manages logic. Confusing them is the fastest way to induce "Report Fatigue"—that dread users feel when a dashboard looks like a spreadsheet with buttons.

The Economics of Pixel Real Estate

Think of your report canvas like the cockpit of a high-performance aircraft. You have limited space before the controls become overwhelming.

- **The Slicer (The Joystick):** These are your primary flight controls. They belong on the canvas because the user needs them constantly. If a Regional Manager needs to toggle between "Store A" and "Store B" twenty times a minute to spot margin variances, that toggle must be a Slicer. It costs you screen space, but it buys you **zero-friction interaction**.
- **The Filter Pane (The Circuit Breakers):** These are critical controls that shouldn't be touched during normal flight. This right-hand pane is for "sanitation" and "scope." Use this to quietly remove null values, exclude internal test accounts, or hard-code the fiscal year. Keep the canvas clean for the visuals that actually matter.

The Golden Rule:

New Power BI developers often suffer from "Slicer Gluttony." They plaster slicers for *everything* on the canvas—Year, Month, Region, SKU, Salesperson, Warehouse.

The result is a dashboard where 40% of the screen is admin buttons and only 60% is insight. If a user only changes a filter once a month, banish it to the Filter Pane.

2. The Hierarchy of Scope (The Filter Pane)

When you drag a field into the Filter Pane, you aren't just hiding data. You are establishing the "Rules of Engagement" for your report. You are operating on a strict hierarchy.

Failing to understand this leads to dangerous reporting errors—like a user thinking they are looking at global revenue when they are unknowingly restricted to a single sales channel.

- **Visual Level (The Surgical Knife):** This applies logic *only* to the specific chart you have selected. Use this to correct context-specific data anomalies.
- *The SmartGear Scenario:* You are building a bar chart tracking "Logistics: Units Shipped." However, your sales data includes non-physical items like **"Extended Warranty Plans"** and **"Digital Software Downloads."** These have revenue, but they don't go on a truck. You apply a Visual Level filter to exclude `Product Category = 'Virtual'` *only* for this chart. The "Total Revenue" card next to it remains accurate and inclusive, while your logistics chart reflects reality.
- **Page Level (The Local Law):** This applies to every visual on the current tab.
- *The SmartGear Scenario:* You are designing a tab dedicated to "Wearables Performance." Drop `Category = 'Smartwatches'` into the Page Level filters. Now, every chart on that page automatically reflects that category. It saves you from repetitive setup and ensures no stray data from "Home Audio" leaks into the view.
- **Report Level (The Nuclear Option):** This applies to every visual, on every tab, in the entire file.
- *The SmartGear Scenario:* Use this for global sanitation. Your IT department frequently creates dummy orders to test the checkout system, logged under `Customer = 'Internal_Test_Admin'`. Or perhaps you have legacy data from a pre-2020 ERP system that is no longer comparable. Nuke these at the Report Level. This ensures that no matter where the CEO clicks, that bad data never resurfaces.

3. The Art of the Slicer (User-Facing Controls)

Your end users are "Smart Beginners"—professionals who understand retail KPIs perfectly but have zero patience for bad software design. Your slicers act as the bridge between their intent and the data. Configuration is not about aesthetics; it is about cognitive ease.

List vs. Dropdown: Managing Friction

Hick's Law states that the time it takes to make a decision increases with the number of choices. apply this to your UI:

- **The List:** If you have fewer than 8 items (e.g., Sales Regions), use a "List" (open checkboxes). It allows the user to see all options immediately without a click. It invites exploration.
- **The Dropdown:** If you have 50 Store Locations, a list will clutter your screen. Use a Dropdown. It saves space, but remember it adds "click friction" (open, scroll, select, close). Use it only when necessary.

The Date Slicer Trap

By default, Power BI creates a "Between" slider for dates (a horizontal bar with two drag handles).

- **The Problem:** In a high-velocity retail environment, nobody has the time (or motor precision) to drag a tiny handle exactly to "November 1st." It is slow and frustrating.
- **The Fix:** Switch the slicer type to **"Relative Date"**.
- **The Impact:** This allows you to set logic like **"Last 30 Days"** or **"This Week."** This creates a "rolling" report. When a SmartGear manager opens the report on Monday morning, it automatically updates to show the latest window of data. You just saved them—and yourself—a manual adjustment.

DAY 5

Control Freak: Single vs. Multi-Select

Excel users are trained to use `CTRL + Click` for multi-selection. This is a trap in a touchscreen or dashboard environment. Users forget to hold CTRL, click a new item, and accidentally reset their entire view.

- **The Professional Setup:** Go to Slicer Settings > Selection. Enable **"Select All"** (a panic button for users to reset the view). For metrics that cannot be aggregated—like "Exchange Rate" or "Inventory Turnover Ratio"—force **"Single Select."** You cannot sum the Euro rate and the Dollar rate; the user *must* pick one.

4. Advanced Control: Sync Slicers & Interactions

This section distinguishes the Excel user from the Power BI Developer. It addresses two specific behaviors that cause users to lose trust in a report.

A. Sync Slicers (Curing Digital Amnesia)

Imagine walking from your living room to your kitchen, and suddenly forgetting what year it is. That is the experience of a user who selects "2024" on Page 1, clicks to Page 2, and sees data for "All Years."

- **The Fix:** You must explicitly link your slicers. Open the **Sync Slicers Pane** (`View -> Sync Slicers`). You will see a matrix of your report pages.
- *Check "Sync":* This passes the filter context (e.g., Year = 2024) to the other pages.
- *Check "Visible":* This decides if the physical slicer box appears on that page.

This creates **Object Permanence**. The user feels like they are navigating a single, cohesive application, not a binder of disjointed spreadsheets.

B. Visual Interactions (Filter vs. Highlight)

This is one of the most polarizing default behaviors in Power BI. When you click a bar in a chart (e.g., "Headphones"), what happens to the other charts on the page?

- **The Default (Highlighting):** The other charts dim the non-selected data (turning them into semi-transparent "ghost bars") and highlight the "Headphones" portion.
- *The Issue:* In a Line Chart showing a 12-month trend, highlighting just a sliver of the line makes it unreadable. It creates visual noise rather than clarity.
- **The Professional Choice (Filtering):** You want the other charts to physically *remove* the unrelated data and rescale the axes to focus entirely on "Headphones."
- **Execution:** Select a visual, go to the `Format` tab, and click `Edit Interactions`. You will see small icons appear above every other chart. Toggle from the "Highlighter" icon to the "Filter" funnel icon.

The Bottom Line: Highlighting is useful for comparing a part to the whole (e.g., "How much of our total revenue came from Headphones?"). But for deep analysis, Filtering is almost always the superior choice because it removes distraction.

5.3 Conditional Formatting: highlighting Key Performance Indicators

From Static Painting to Dynamic Diagnostics

Let's be honest about the old workflow. In Excel, you see a low number, you manually fill the cell with a red background, and you feel productive. But that is not analysis; that is painting. And it is a liability.

That red cell stays red until you remember to scrub it. If the data refreshes next month and performance improves, your static formatting is now lying to you.

In Power BI, conditional formatting is a **diagnostic system**. It acts as the nervous system of your data model. You define the rules of "pain" (margin erosion) and "pleasure" (growth), and the report reacts automatically. For the SmartGear analyst, this shifts your role entirely. You stop being the person pointing at the problem and become the architect of a system that flags the problem *for* you. It directs the stakeholder's eye immediately to the outliers—the gaming consoles losing money or the accessories driving profit—rather than forcing them to scan through the noise of averages.

Accessing the Logic (The Hidden `fx` Button)

Here is a major friction point for new users: Microsoft does not put conditional formatting on the main ribbon. They buried the logic deep inside the formatting pane. To activate it, you have to know the secret handshake.

1. Select your visual.
2. Go to the **Visualizations pane** and click the **Format visual** icon (the paintbrush).
3. **For Tables/Matrices:** Navigate to **Cell elements**. You will see toggles for *Background color*, *Font color*, *Data bars*, and *Icons*.
4. **The Critical Step:** Simply turning the toggle "On" applies default logic, which is rarely what you want. To control the rules, you must click the small **`fx` (Function)** button next to the toggle.

For Charts (Bar/Column): It is even more obscure. Go to **Columns** (or Bars) -> **Color**. You won't even see a toggle. You have to instinctively know to click the **`fx`** icon next to the default color swatch.

The Three Logic Engines

Once you click that `fx` button, you are presented with three distinct engines. Choosing the wrong one is the most common reason dashboards fail to tell a coherent story.

DAY 5

1. Gradients (The Heatmap Approach)

This assigns color intensity on a continuous spectrum. Use this for **pattern recognition**—like a heatmap identifying high-density sales regions—where the *concentration* of data matters more than the precise number.

The Analyst's Edge: Always enable **"Diverging"**. Without this, Power BI simply scales from light to dark. With "Diverging," you can set a "Center" value (e.g., 0% variance). This ensures that positive growth is strictly Blue and negative growth is strictly Orange, creating an instant, binary separation of performance.

2. Rules (The KPI Standard & The "Trap")

This is the classic "If/Then" logic. *If Margin < 20%, then Red.* Simple, right? This is where 90% of Excel users crash in Power BI.

The Brutal Truth (The "Percent" Trap): When you open the Rules dialog, the default logic for input is set to **Percent**, not **Number**. If your data ranges from -5% to -10%, the -5% is technically the "top 100%" of that specific view. Your formatting will break the moment you filter the data.

The Fix: Always switch both dropdowns from "Percent" to **"Number"** immediately. You want to evaluate the actual metric (0.20), not its relative position in the visual.

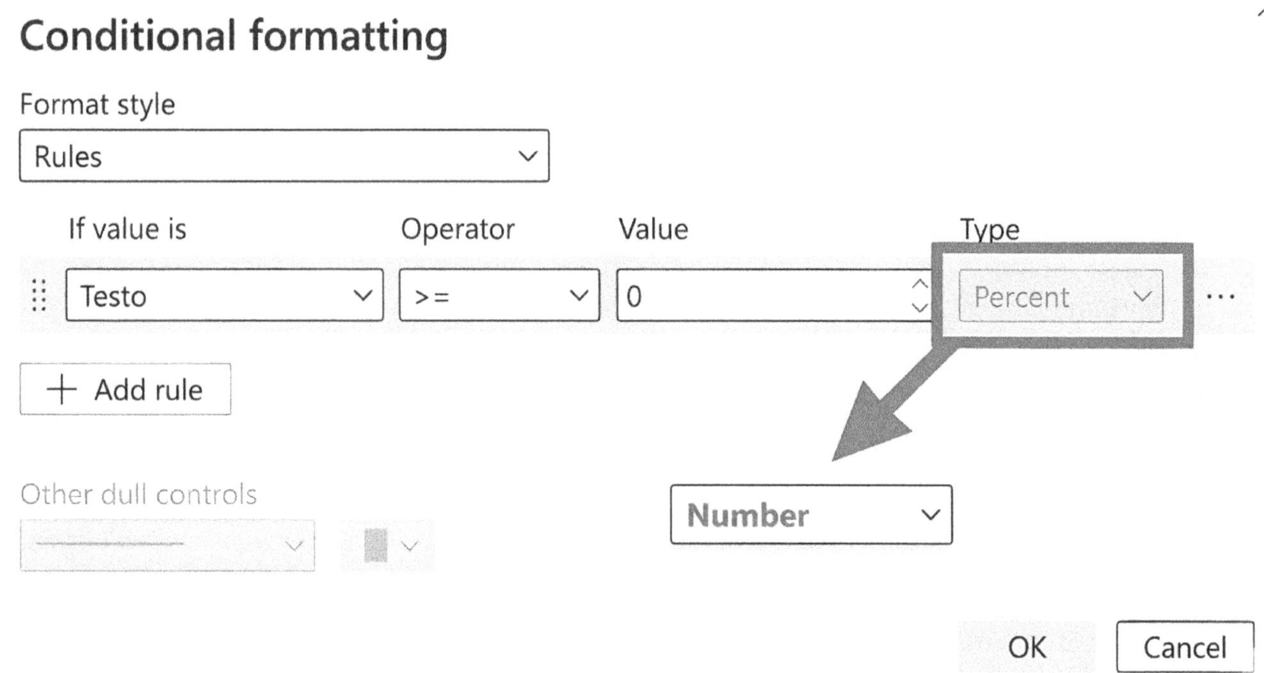

*5.3- **The Default Settings Trap.** When setting up Conditional Formatting rules, Power BI defaults the dropdown to "Percent." This does not mean "format my percentage measure." It means "calculate the relative 0-100% distribution of the data currently on screen." If you leave this as Percent, your colors will change wildly based on slicers. You must manually change this dropdown to Number for every single rule.*

3. Field Value (The Scalable Choice)

Instead of setting rules inside the visual, you select a DAX measure that outputs a specific color code. Why? **Brand Consistency and Scalability.**

SmartGear has specific corporate colors. You don't want a random "Excel Green"; you want *SmartGear Emerald*. If you rely on the **Rules** engine, and Marketing changes the official red from `#FF0000` to `#C0392B`, you have to manually update 15 different charts. That is wasted time.

The Solution: Create a dedicated DAX measure:

`KPI Color = IF([Profit Margin] < 0.18, "#C0392B", "#27AE60")`

Select **Field Value** in the formatting dialog and point it to this measure. Now, if the VP of Finance decides that 18% is no longer the target and raises it to 20%, you update *one* measure, and the entire report suite updates instantly.

Visualization Specifics: Finding the "Empty Calories"

Different visuals require different psychological approaches. For SmartGear, we use charts to expose the truth behind revenue.

The "Empty Calories" Chart

High sales do not always mean high success. In retail, we often see products that generate massive revenue but zero profit—we call these "Empty Calories."

- **The Setup:** Create a Clustered Column Chart. Set the **Y-Axis (Height)** to *Total Sales*.
- **The Formatting:** Use Conditional Formatting on the **Columns** -> **Color**. Base it on *Profit Margin* (Red for low, Green for high).
- **The Insight:** You will likely see the **Gaming Consoles** bar towering over the others—it is huge because revenue is high. But if the bar is bright **Red**, you instantly know that despite the volume, we are bleeding margin on every unit sold. Meanwhile, a short bar for **HDMI Cables** might be bright Green. This visual tells you where to focus your cross-selling efforts immediately.

Tables & Matrices:

- **Data Bars:** Check the **"Show bar only"** box. This hides the text number. Use this when the *magnitude* is more important than the *precision*. It cleans up the table significantly.
- **Icons:** Arrows and flags communicate "Good/Bad" faster than colors because they use shape *and* color simultaneously.

The Integrity of Design: Cognitive Load & Accessibility

Just because you *can* format everything, doesn't mean you *should*.

The "Clown Vomit" Principle

If everything is highlighted, nothing is highlighted. A table covered in red, yellow, and green squares looks like "clown vomit" (yes, that is the standard industry term for over-formatted reports). It induces anxiety, not insight.

Rule of Thumb: Highlight only the exception. If a number is "normal," leave it white or transparent. Only color the disasters or the miracles.

The Accessibility Imperative

Approximately 8% of men (statistically including some of your C-level stakeholders) have Color Vision Deficiency (CVD). If you rely solely on a "Red = Bad, Green = Good" background, a color-blind executive sees two shades of muddy brown. You are effectively hiding data from the person signing your paycheck.

The Professional Solution:

- **Use Icons:** An "Up Arrow" is universally understood, regardless of color.
- **Blue/Orange Palette:** Use Blue for good and Orange for bad. These are distinguishable by almost all forms of color blindness.
- **High Contrast:** Ensure your font color (Dark Grey/Black) stands out against your background color.

5.4 Drill-Through and Tooltips: Adding Depth without Clutter

Stop trying to cram the entire history of a company onto a single 16:9 canvas. It looks like a hoarder's living room. It's messy. It's useless.

In cognitive psychology, we deal with **Cognitive Load Theory**. The human working memory has a hard cap. If you throw a 5,000-row table, three pie charts, and six different KPIs at a manager simultaneously, you aren't informing them. You are paralyzing them.

The antidote is **Progressive Disclosure**.

Think about Google Maps. When you look at the entire United States, you don't see every alleyway in Chicago. You see states. You see highways. You only see street names when you zoom in. In Power BI, we replicate this natural behavior using two specific mechanisms: **Tooltips** (The Hover) and **Drill-Through** (The Click).

Tooltips: Context on Demand (The "X-Ray")

Think of your main dashboard as the surface of the ocean. It shows the waves, the general direction, the storm clouds. Tooltips are your snorkeling gear; they allow you to peek just beneath the surface without needing a submarine crew.

The Default vs. The Professional

Out of the box, Power BI is lazy. It gives you a standard black text box when you hover over a chart. It repeats exactly what the user can already see: "Country: USA, Sales: $5M."

DAY 5

This is redundant clutter.

The "High Efficiency" approach is the **Visual Tooltip** (Report Page Tooltip). Instead of static text, you embed a miniature, fully functional report page inside that popup.

- **The Experience:** Imagine hovering your mouse over a bar chart for "France." Instantly, a small window materializes. Inside isn't just a number; it's a micro-trend line of French performance over the last 12 months and a breakdown of the top 3 products.
- **The Logic:** You are answering the user's next logical question—*"Is France growing or shrinking?"*—without forcing them to click, filter, or leave the current view.

The Brutal Truth About Performance

Here is the technical reality that most tutorials gloss over: **Tooltips calculate on the fly.** Every time your mouse moves from one bar to the next, Power BI fires a query to the engine to render that mini-report. If you put a complex matrix with 15 heavy DAX measures inside a tooltip, your report will feel "sticky." The cursor will lag. It ruins the experience.

- **The Rule:** Keep tooltips lightweight. They are for quick insights, not deep calculus.

Drill-Through: The Forensic Deep Dive (The "Click")

If Tooltips are snorkeling, Drill-Through is the submarine dive to the bottom of the Mariana Trench. This feature bridges the gap between the polished "Executive Summary" and the ugly "Excel Grid." As an analyst, you deal with stakeholders who constantly say, "I just want to see the data." They feel insecure without the raw rows. Drill-Through is how you give them that security blanket without destroying your dashboard's usability.

The Workflow

1. **The Trigger:** A user sees a red flag. For example, the "East Coast Region" bar is significantly below target.

2. **The Action:** They right-click that red bar and select "Drill Through."

3. **The Result:** They are transported to a completely different page—the **Target Page**.

The Target Page Strategy

This page is designed for forensic analysis. This is the one place where it is acceptable to have a detailed table with Invoice IDs, Dates, Customer Names, and exact margins. Because the user arrived here via the "East Coast" bar, this detailed table is *already filtered* to only show East Coast transactions. You have guided them from the "What" (Sales are down) to the "Why" (These specific 5 invoices were refunded).

The UX Reality Check: Right-Click vs. Buttons

Here is the friction point where logic meets human behavior.

The Problem: Casual users do not right-click.

If you deliver a report to a Sales Director relying on them right-clicking a bar to find the details, they will likely call you to ask, "Where is the data?" Furthermore, right-clicking is clumsy on touchscreens and iPads, which is exactly where many executives consume data.

The Solution: Drill-Through Buttons

You must make the affordance explicit.

1. **Create a Button:** Label it "See Details."

2. **The Logic:** Configure the button so it is greyed out (disabled) by default.

3. **The Interaction:** When the user clicks a specific bar on your chart, the button suddenly lights up (enables). You can even use DAX to make the button text dynamic: *"See Details for [Selected City]"*.

This satisfies the "High Efficiency/Low Friction" requirement. It guides the user's hand, making the path to the data obvious and clickable.

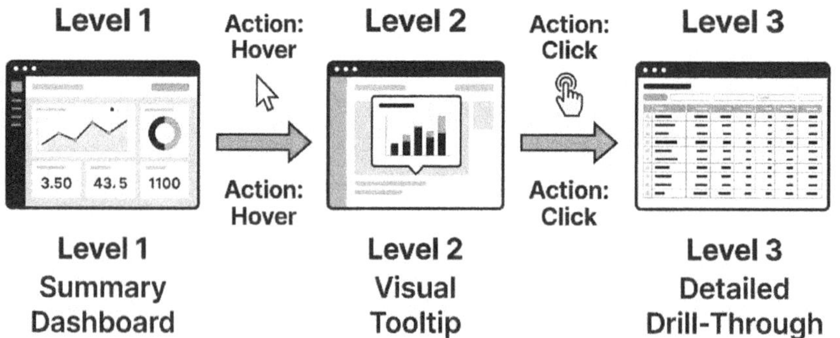

*5.4 - **Progressive Disclosure Architecture.** Do not force every column of data onto the home page. The executive summary (Level 1) should be clean and distraction-free. By using interactive elements—like a Hover to reveal a Tooltip (Level 2), and a Click to jump to a Drill-Through page (Level 3)—you guide the user's curiosity without overwhelming them with a wall of numbers.*

5.5 Layout and Design: Principles for a Clean User Interface

The Canvas: Finite Space vs. Infinite Scroll

Here is the first psychological hurdle you must clear: **Excel is a scroll; Power BI is a slide.** In Excel, your thinking is vertical. You are conditioned to an infinite grid where you can always add "just one more row" or scroll down endlessly to find a grand total. In Power BI, this habit is destructive. Power BI operates on a **finite canvas**. This is not a limitation of the software; it is a constraint designed to force prioritization. Decision-makers cannot compare two metrics if one is on the screen and the other is three scrolls down. The default aspect ratio is **16:9** (typically 1280x720 or 1920x1080 pixels). This isn't arbitrary. It mirrors modern monitors and boardroom projectors.

DAY 5

The Brutal Truth: If your user has to scroll to see the "bottom line," you have failed. You haven't built a dashboard; you've built a web page. If you feel the urge to cram more data onto the page, don't expand the canvas size—create a second page or a drill-through.

- **Action:** In Power BI Desktop, immediately enable **"Snap to Grid"** (View > Show Gridlines / Snap to grid).
- **The Experience:** Without this, your dashboard will suffer from "visual vibration"—that subtle, subconscious anxiety users feel when elements are 2 pixels misaligned. It looks unprofessional. Worse, it makes the data appear strictly less trustworthy.

The "Golden Quarter" and Eye-Scanning Patterns

Users do not "read" dashboards like a book; they scan them like a billboard. Research by the **Nielsen Norman Group** proves that on data-heavy screens, the human eye follows a specific trajectory known as the **Z-Pattern**. You must align your data with biology, not fight against it.

1. **The Priority Zone (Top-Left):** This is the "Golden Quarter." It is the most valuable real estate on your screen.

- *The Mistake:* Many beginners put their company logo or a decorative title here.
- *The Fix:* Put your "Big Angry Numbers" (BANs) here. Total Revenue, Net Margin, or the single most critical KPI that defines success or failure. If the user only looks at this corner and walks away, they should still know the health of the business.

2. **The Context Zone (Top-Right):** As the eye scans right, it looks for context. This is where your Slicers (Date, Region, Segment) belong.

3. **The Terminal Area (Bottom-Right):** The scan ends here. This is the "graveyard" for attention. Use this area for granular tables or detailed matrices—data that is only needed if the top-left numbers trigger an alarm.

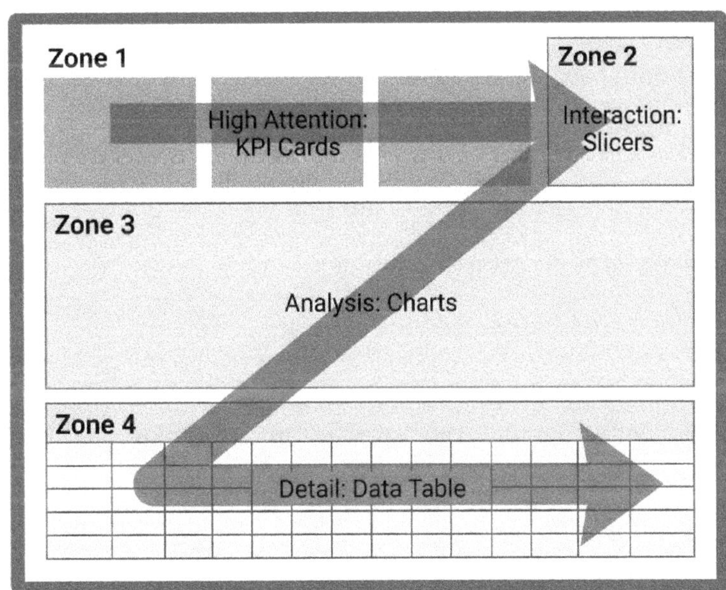

*5.5 - **The Z-Pattern Architecture.** Western readers scan a page from top-left to bottom-right in a "Z" shape. Place your most critical, immediate insights (KPI Cards) in Zone 1. Place the interactive controls (Slicers) at the end of that first horizontal scan (Zone 2). The middle (Zone 3) holds the comparative charts, and the bottom (Zone 4) is reserved for the dense, granular tables—the exact opposite of how an Excel spreadsheet is built.*

Managing Cognitive Load with Gestalt Principles

Your brain burns glucose to process visual information. Every time a user has to guess which chart relates to which filter, they burn energy. If the "cognitive load" is too high, they will close the report and ask you to "just send it in Excel."To prevent this, we use **Gestalt Principles**—psychological laws describing how humans perceive order in chaos.

- **Law of Proximity:** Things that are close together are perceived as related.
- *The Excel Habit:* Scattering pivot tables wherever they fit.
- *The Power BI Rule:* If a "Sales by Region" chart is controlled by a specific slicer, place them adjacent to each other. Do not make the eye jump back and forth across the screen.

DAY 5

- **Law of Common Region (Enclosure):** Objects within a boundary are seen as a group.
- *The Fix:* Instead of using heavy black borders around every single chart (which looks like a prison cell), use a subtle, light-grey background shape to visually "hold" a set of related metrics together. This creates an invisible container that organizes the chaos without adding "ink."

The Data-Ink Ratio: The Art of Subtraction

Edward Tufte, the godfather of data visualization, coined the term **Data-Ink Ratio**. The principle is ruthless: *Every pixel on the screen should serve the data. If it doesn't, delete it.* In Excel, we are used to gridlines, borders, and bold headers to separate cells. In a visual dashboard, these are just noise.

- **Kill the Gridlines:** On a bar chart, if you have data labels on the ends of the bars, you do not need a Y-axis, and you certainly do not need horizontal gridlines. They cut through the data and add visual friction.
- **Ban 3D Effects:** Never use 3D pie charts or shadowed bars. They distort the physical area of the graphic, making the data harder to read for the sake of looking "fancy."
- **The "Clean Up" Routine:**

1. Turn off background visuals for individual charts (let them float on the canvas).
2. Lighten your axis text colors (grey is better than black).
3. Remove chart borders. Let the whitespace define the boundaries.

The Reality Check: A "clean" interface isn't about minimalism for art's sake. It's about efficiency. A cluttered dashboard hides insights; a clean dashboard screams them. You are no longer just reporting numbers; you are designing the user's attention.

Day 6 - Publishing and Sharing: The Collaborative Workflow

6.1 Publishing to Power BI Service: The Handoff

The Great Divide: Decomposition and Deployment

Clicking "Publish" is the definitive moment in the Power BI workflow. It is the bridge between the private, chaotic laboratory of your desktop and the public, governed showroom of the Cloud.

In your Excel life, "sharing" is pedestrian. You attach a workbook to an email or drop it into a SharePoint folder. The file remains a single, monolithic block containing your data, your logic, and your charts. Power BI is different. When you press **Publish**, you aren't just copying a file. You are triggering a **decomposition process**.

You are deploying software.

1. The Cloud Decomposition: Anatomy of a Split

Why does Power BI rip your file apart? Efficiency. In a local `.pbix` file, the data and the visuals are fused. But in a cloud architecture, these two elements have distinct, competing needs. Data demands raw processing power (RAM and CPU) to crunch numbers. Visuals need lightweight rendering code to load quickly in a web browser.

By splitting them, Power BI allows you to reuse one heavy dataset for twenty different reports without duplicating the load.

Think of it like a movie theater.

- **The Semantic Model (formerly Dataset)** is the **Film Reel** in the projection room. It is heavy, it contains all the frames (data), and it follows a specific sequence (logic).
- **The Report** is the **Projector**. It is merely light and lenses. It reads the reel and displays it on the screen.

When you hit Publish, the Service performs surgery. It uploads your `.pbix`, extracts the "Film Reel" to a backend server, and places the "Projector" in the frontend gallery. They are now two separate living entities, linked only by an invisible wire.

The Confusion: This is where Excel users get tripped up. You will log into the Service and see two icons with the *exact same name*. One is blue (Report), one is orange (Semantic Model).

If you delete the orange one, the blue one breaks instantly—it becomes a projector with no film. If you delete the blue one, the data remains safe. Excel users rarely think about "dependencies," but in Power BI, you must.

2. The Destination Hierarchy: Where Data Lives

Before the upload completes, you face a critical choice: **Where does this go?** The dialog box asks for a destination, and making the wrong choice here can paralyze your operations later.

My Workspace: The Personal Sandbox

This is the default selection. It feels safe. It looks like "My Documents." **It is a trap.**

"My Workspace" is tied strictly to your individual email account. If you build a mission-critical financial report here and then leave the company—or even just go on vacation—that report is locked in your digital vault. No one else can edit the backend. It is an operational dead-end. Use this strictly for scratchpad testing.

Collaborative Workspaces: The Team Hub

These are shared environments (e.g., "Finance North America").

They are independent of any single employee, acting like shared folders with specific roles (Admin, Member, Contributor).

The Licensing Reality Check: Here is the "paywall" moment Microsoft doesn't advertise on the download page. You can use Power BI Desktop for free forever.

You can publish to My Workspace for free. **But you cannot share data privately in a Team Workspace without a Pro License.**

If you are the "Excel Guy" trying to modernize your department on a zero budget, this is where you hit the wall. Sharing costs money.

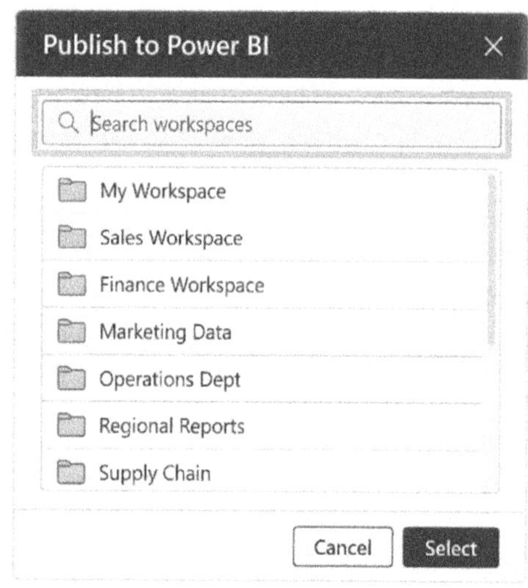

*6.1 - **Search, Don't Scroll.** When your company scales, you will have access to dozens of workspaces. Never scroll and click blindly; accidentally publishing your "Test" report into the "Executive Finance" workspace will overwrite live data. Always use the Search Bar (highlighted) to isolate your exact destination before clicking Select.*

DAY 6

3. The "Push" Protocol

The physical act of publishing is simple, but it requires discipline.

1. **Authentication:** You must be signed into the Service via the **Desktop app.**
2. **The Save Mandate: Power BI will force a local save before publishing.**
3. **The Selection: Upon clicking Publish, a dialog box appears. This is your routing switchboard.**

4. The "Overwrite" Trap and Impact Analysis

Power BI operates on unique identifiers (GUIDs). When you publish a file named `Q3_Results.pbix` to a workspace that already has that file, the Service recognizes the ID match. It assumes you are pushing an update (Version 1.1) and prepares to wipe the old version (Version 1.0) to replace it with yours.

There is no `Ctrl+Z` in the Cloud.

If you accidentally publish an older version of a file over a newer one in the Service, the newer one is gone. Forever. Unlike SharePoint or OneDrive, Power BI Service does not inherently keep a "Version History" stack for datasets. Your local file management is your *only* safety net.

Furthermore, if you have been smart and built five different reports connected to that single Semantic Model, publishing changes to the model puts them all at risk. This is why the **Impact Analysis** warning is the most important pop-up you will see. It isn't just a "Are you sure?" box. It is a blast radius map. It tells you: *"If you update this dataset, you are about to affect 4 other dashboards owned by Steve in Accounting and Sarah in Marketing."*

5. Post-Publishing Architecture: The Golden Rule

Once the upload bar hits 100% and you click the link to open your report in the browser, you have crossed a threshold. You are now in the "Consumption Layer."

You will notice an "Edit" button in the browser. **Do not touch it.**

This is the most common mistake efficient professionals make. They see a typo in the browser and think, *"I'll just fix it here quickly."* If you change the report in the Cloud, your local Desktop file—the master source—is now outdated. The next time you publish from Desktop to Cloud, your local file will overwrite the Cloud version, and that "quick fix" you made in the browser will be obliterated.

Treat your Power BI Desktop file (`.pbix`) as the **Master Source Code**. All changes—even moving a text box one pixel to the left—must happen locally and be republished. The flow is always One-Way: **Desktop $\rightarrow$ Cloud**. Never the reverse.

DAY 6

6.2 Workspaces vs. Apps: Structuring Content for Teams

The Two Environments: Kitchen vs. Dining Room

Up to this point, you have been operating exclusively in the **Workspace**. In the professional Power BI workflow, the Workspace is the **Kitchen**.

Picture a commercial kitchen during the Friday night dinner rush. It is hot. It is noisy. It is chaotic. There are raw ingredients (datasets) scattered on stainless steel counters, sauces reducing (dataflows), and half-plated dishes (reports) waiting for a final garnish. This is where the cooks—you and your fellow analysts—work. It is a place of creation, trial, and error.

Do not invite your customers into the kitchen.

If you share a report directly from a Workspace, you are essentially inviting your CEO to eat off the prep table. They might see a visual you are halfway through fixing. They might get confused by a file named `Sales_v3_FINAL_TEST_DO_NOT_TOUCH`. Worse, they might accidentally move a slicer, panic, and claim they "broke the numbers."

Instead, you serve them in the **App**.

The App is the **Dining Room**. It is curated, branded, air-conditioned, and quiet. The navigation is simplified into a neat menu. Most importantly, the dining room is **read-only**. The customer can consume the meal, but they cannot walk back into the kitchen and change the recipe.

The Workspace: Collaboration Only

The Workspace exists for one reason: to build the content. It is a collaborative environment, not a consumption environment. To maintain order in the kitchen, you must strictly manage who holds the keys. Here is the brutal truth about permissions: **Most organizations are too generous.**

They default to making everyone an "Admin" or "Member" because it is convenient. This is a security nightmare waiting to happen. If everyone is an Admin, everyone has the power to delete the entire workspace by accident. Do not let this happen.

- **Admin:** God mode. They can update, delete, and add other users. **Limit this to 1-2 people** (Lead Analyst and a backup).
- **Member:** The "Head Chef." Can publish Apps and share items, but cannot delete the workspace. **Limit to Senior Analysts.**
- **Contributor:** The "Line Cook." Can edit reports and datasets but *cannot* push the final "Publish App" button. **Warning:** Because they can edit, they can see *all* data in the model. RLS does not apply to them.
- **Viewer:** The "Health Inspector." This is the **only** role in the Workspace where Row-Level Security (RLS) is active. Use this role to test what a restricted user sees before going live.

DAY 6

⚠ CRITICAL SECURITY PROTOCOL: The RLS Bypass Rule

This is where 90% of new administrators accidentally leak data.

If a user has permission to EDIT content (Admin, Member, or Contributor), Power BI ignores Row-Level Security.

The logic is simple: You cannot cook a meal if you are forbidden from seeing the ingredients. If you make a Regional Sales Manager a "Contributor" so they can fix a typo in a report, **they will instantly see global data**, bypassing any filters you wrote to restrict them to their region.

- **Rule of Thumb:** If they need to see restricted data (RLS), they must be a **Viewer** or an **App Consumer**. Never a Contributor.

The App: The Consumption Layer

An **App** is a packaged bundle of reports and dashboards from a single Workspace. While Excel users are accustomed to emailing a file—where the creation file and the consumption file are identical—Power BI separates them. This separation offers three massive advantages.

First, **Decoupling (The "Staging" Environment).** This is the most critical feature for your sanity. You can make massive changes to a report in the Workspace—breaking visuals, testing new measures, rearranging pages—and the users in the App **will not see a thing** until you hit "Update App." You can finally work on a live project without fear of someone calling you to ask why the charts look weird.

Second, **Curated Navigation.** In a workspace, files are listed alphabetically or by date, which is useless for an end user. In an App, you control the experience. You can create sections like "Executive Summary," "Regional Detail," and "Raw Data," regardless of which actual report file those pages come from.

Third, **Permissions Separation.** The list of people who build the data (Workspace access) and the people who read the data (App access) are two completely different lists. This keeps your development secure and your distribution broad.

The "Audiences" Feature: One App, Multiple Views

Historically, Power BI had a frustrating limitation: "One Workspace = One App."

If Sales needed Report A, and Finance needed Report B, you often had to duplicate the dataset into two different workspaces just to keep them separate. This created "data silos" and maintenance hell. You fix a measure in one file, but forget the other. Numbers drift. Trust evaporates.

Since late 2022, Power BI introduced **Multiple Audiences**. This allows you to partition your Dining Room into "VIP Sections." You can now have a single source of truth (one workspace) but distribute it differently to different groups. Consider this scenario: You have a "Company Performance" Workspace containing a *Sales Report*, a *Marketing Report*, and a *Salary Audit*.

- **Audience 1 (All Staff):** You create a view that includes only *Sales* and *Marketing*.
- **Audience 2 (HR Only):** You create a view that includes the *Salary Audit*.

DAY 6

The Critic's Warning: Do not confuse **Audiences** with **Security**.

Hiding a report via Audiences is like putting a "Do Not Enter" sign on a door. It hides the door from the hallway, but it doesn't lock it. If a clever user gets the direct URL to the hidden report, they might still access it.

- **Audiences** control *visibility* (UI/UX).
- **Row-Level Security (RLS)** controls *data access* (Logic).

If you need to ensure that a Sales Manager sees *only* their region's data within a shared report, you must use RLS inside the dataset logic. Audiences are for managing clutter, not for managing high-stakes secrets.

The Publishing Workflow Checklist

Moving from Desktop to a deployed App is not just "saving a file." It is a deployment pipeline. Professionals follow a rigid sequence to ensure no errors reach the Dining Room.

1. **Desktop:** Save your `.pbix` file. Run the "Impact Analysis" check to see what downstream dashboards might be affected.

2. **Desktop:** Publish to **Workspace** (The Kitchen).

3. **Service (Workspace):** *Stop.* Do not publish the App yet. Verify the data refresh. Check that the visuals rendered correctly on the web (fonts and spacing sometimes shift from Desktop to Web).

4. **Service (Workspace):** Click **"Create App"** (or "Update App").

5. **Service (App Menu):**

    - **Setup:** Name your App clearly. Add a support link (so they email a helpdesk, not you directly).
    - **Content:** Select which reports to include.
    - **Audience:** Define who sees what. Double-check your distribution lists.

6. **Service:** Click **Publish App**.

Once published, you train your users to bookmark the **App URL**. They should never need to see the workspace link.

The Financial Reality Check:

There is always a catch. To view a Power BI App, the consumer must have a license.

- **Scenario A:** You have **Power BI Pro** licenses for everyone. Great, everyone can view the App.
- **Scenario B:** You have **Premium Capacity** (or Fabric Capacity). This is the "Open Bar." You pay a high flat rate for the capacity, and "Free" users can view the Apps hosted there.

The Trap: If you do *not* have Premium Capacity, you cannot share an App with a "Free" user. Microsoft will prompt them to start a Pro trial (and eventually pay) the moment they click your link. Be aware of your organization's licensing model before you promise a dashboard to 500 unlicensed users.

6.3 Automating Success: Configuring Scheduled Refreshes

You have just hit "Publish." The visuals look crisp. The insights are compelling. You close your laptop, satisfied with a job well done. But here is the invisible problem.

The moment that report landed in the Cloud, it started dying. When you publish to Power BI, you are not piping a live stream of data; you are uploading a **snapshot**.

It is a static memory of what your Excel file or SQL database looked like at that exact second. If you walk away now, your users will open that dashboard next Tuesday and see *last* Tuesday's numbers. There is no quicker way to incinerate your credibility than a CEO asking, *"Why does this say Sales are zero for yesterday?"*

To keep the report alive, you must configure the **Scheduled Refresh**. This is the heartbeat of your dashboard. It automates that tedious "refresh-save-republish" loop you used to suffer through in Excel.

1. The Bridge: Understanding Data Gateways

Before you can set a schedule, the Power BI Service (which lives in Microsoft's Azure Cloud) needs a way to reach your data.

Scenario A: Cloud-Native Sources (e.g., SharePoint Online, Salesforce).

This is usually frictionless. Both your report and your data live on the internet. Power BI can simply "reach out and touch" the data using the credentials you provide. No extra software required.

Scenario B: On-Premises Sources (e.g., Local SQL Server, an Excel file on a mapped network drive).

This is where it gets complicated. The Cloud cannot see inside your company's private network. Your corporate firewall is designed specifically to stop external traffic—including Microsoft—from peeking into your servers.

You need a bridge. We call this the **On-premises Data Gateway**.

The Brutal Truth About Gateways:

If you work in a large corporate environment, do not try to be a hero. You likely cannot install a Gateway yourself because you don't have admin rights to the servers. You will encounter "Organizational Friction." You must ask IT to install the **Standard Gateway**.

- **Standard Mode (The Professional Choice):** It runs on a server that never sleeps. It supports all connection types and serves multiple developers.
- **Personal Mode (The Trap):** You can install this on your own laptop. It sounds tempting. **Don't do it.** Why? If you close your laptop lid, the refresh fails. If you go on vacation, the refresh fails. If your Wi-Fi drops, the refresh fails. Use Personal Mode only for testing prototypes. Never use it for production.

DAY 6

2. The Configuration Workflow

Once the bridge is built, you must tell the report how to cross it. Navigate to your Workspace, hover over the **Semantic Model** (Dataset), click the **ellipsis (...)**, and select **Settings**.

Phase 1: The "Drive Letter" Fallacy

This is the single most common error for Excel users moving to Power BI.

In Power BI Desktop, you might have connected to `Z:\Finance\Budget.xlsx`. But the Gateway server doesn't know what "Z:" is. That is just a shortcut on *your* specific computer.

The Fix: You must use the **UNC Path** (e.g., `\\ServerName\Finance\Budget.xlsx`). If the path in the Service doesn't match the path in your PBIX file *character-for-character*, the Gateway will say "Not Configured."

Phase 2: The Handshake

Even if you are logged in on your Desktop, the Cloud service is a separate entity. It needs its own set of keys.

- **Authentication:** You will be asked to sign in again. For cloud sources, this is usually **OAuth2**. For on-prem, it is often **Windows**.
- **Privacy Levels:** You will see options for *Private, Organizational,* and *Public*. Set your internal company data to **Organizational**. This tells Power BI: "This data is safe to mix with other data from within my company, but don't send it outside."

Phase 3: Beating the Traffic

Toggle **"Keep your data up to date"** to On. Now you pick your times.

Here is a secret about cloud computing: it is not infinite. It is a physical server farm shared by millions of people.

- **The Rookie Move:** Scheduling your refresh for 8:00 AM or 9:00 AM.
- **The Reality:** Half the world schedules their reports for the top of the hour. This creates massive resource contention. Your refresh might sit in a queue for 20 minutes before it even starts.
- **The Pro Tip:** Schedule your refresh for **7:45 AM** or **8:15 AM**. By offsetting your time by 15 minutes, you bypass the digital traffic jam and your data loads faster.

3. The Limits (Pro vs. Premium)

Your ambition is limited by your license. This is a hard technical constraint that no amount of clever coding can bypass.

- **Power BI Pro:** You get **8 refreshes per day**. That is once every 3 hours. For 99% of financial and operational reporting, this is sufficient. If a user tells you they need the P&L updated every 10 minutes, they are confusing "Strategic Reporting" with "Operational Monitoring."
- **Power BI Premium (Fabric):** You get **48 refreshes per day** (every 30 minutes).

The "Real-Time" Illusion:

If you are tracking a stock ticker or a manufacturing assembly line and need *second-by-second* updates, Scheduled Refresh is the wrong tool. You need **DirectQuery**. Do not try to force Scheduled Refresh to do a job it wasn't designed for.

4. Handling Failures (The "0 to 1" Rule)

In the world of data automation, "set it and forget it" is a myth. A refresh *will* eventually fail. Passwords expire, servers go down for maintenance, or a colleague will rename that one critical column in the source Excel file.

- **The Notification Protocol:** By default, Power BI only emails the **Dataset Owner** when a refresh fails. If you are the owner and you are on vacation, the report dies in silence.
- **The Fix:** In the refresh settings, check the box: **"Email these users when the refresh fails."** Add your team's distribution list (e.g., `analytics-helpdesk@company.com`). Do not be the single point of failure.

There is a sharper reason to do this. If your dataset fails to refresh **four consecutive times**, Power BI assumes nobody cares. It disables the schedule entirely to save resources. If you ignore the first three emails, on the fourth day, the dashboard stops updating entirely until you manually log in and re-enable it.

Summary: The goal is high efficiency and low friction. By configuring a robust Gateway connection and a smart schedule (offset from peak hours), you transform your report from a static PDF into a living application that serves your team while you sleep.

6.4 Mobile Layout: optimizing Reports for Phones and Tablets

The C-Suite Reality: Why Pixel-Perfect Isn't Enough

Let's be honest. You have likely spent the last three days obsessing over pixel alignment on a 27-inch 4K monitor. The margins are mathematically perfect. The color palette is sublime. It is a masterpiece of logic and design.

But here is the brutal truth that most analysts ignore: **your stakeholder is rarely sitting at a desk.**

Imagine the scenario. Your CEO is standing at Gate B14, waiting to board a flight to London. They have thirty seconds to check if Q3 targets were hit. They pull out an iPhone. If they open your report and have to perform the "pinch-and-zoom" dance just to read a single revenue figure, you have failed. To them, the report looks broken. They don't see your elegant DAX measures or your sophisticated data modeling; they see a messy, unreadable webpage.

The Analyst's Note: Mobile optimization in Power BI is not "responsive web design." Elements do not float around automatically to fit the screen. It is a deliberate, curated secondary view. You are building a specific experience for a specific device. If you don't build it, Power BI simply shrinks the desktop view to fit a 6-inch screen—which is functionally useless.

Entering the Mobile Studio

Fixing this does not require building a new report or duplicating your dataset. You simply need to change your lens.

In Power BI Desktop, go to the **View** ribbon and click **Mobile Layout**.

The expansive white canvas vanishes. It is replaced by a dark, vertical silhouette of a phone overlaid with a fine grid. On your right, the **Page Visuals** pane appears, acting as a holding pen for every chart, card, and slicer you created for the desktop version.

Many beginners freeze here. They worry that moving a visual on this black canvas will wreck the desktop version they just perfected. Let me set your mind at ease: **This process is non-destructive.**

Think of the Mobile Layout as a mask. If you remove a visual from the mobile canvas, you aren't deleting it from the report; you are simply sending it back to the holding pane. You are arranging a playlist, not rewriting the song.

Strategic Triage: The Art of Subtraction

Here is where your "Excel mindset" becomes a liability. In spreadsheets, we crave density. We want to see every row, every column, every variance. On a mobile device, however, you are fighting a war for pixels. You have approximately 320 points of width. You cannot—and should not—fit a 12-column matrix here.

You must apply the **"0 to 1" Rule**:

- **Desktop is for Exploration:** This is where the user sits with a coffee, drills down, and cross-filters complex data.
- **Mobile is for Status Checks:** This is for the glance. The pulse check.

Triage your data ruthlessly. Ask yourself: *"If the building was on fire, which three numbers would I save?"*

Those numbers (KPI Cards, Sparklines) belong at the very top. Mobile usage is built entirely on vertical scrolling. The "House is on Fire" data belongs above the fold. Detailed bar charts go to the bottom. Complex matrices? Leave them out entirely. If the user needs that level of granularity, they should open their laptop.

DAY 6

Independent Formatting: The Game Changer

Historically, mobile design in Power BI was a nightmare. If you reduced a font size to fit the phone screen, it would shrink on the desktop view too. It forced developers to create duplicate measures just for mobile.

The good news: That era is over.

Power BI now supports **Independent Visual Formatting.** This means you can select a visual on the Mobile Canvas, open the **Format** pane, and modify settings like Font Size or Legend Position.

These changes apply **only** to the mobile view. You can shrink a massive 40pt KPI header down to a crisp 18pt for the phone, and your desktop view remains untouched. This decoupling allows you to optimize readability without compromising the big-screen experience.

The Thumb Test: Navigation and Slicers

The mouse is a precision instrument; the human thumb is a blunt object.

If you place five distinct dropdown slicers on your mobile report, you have already lost 50% of your screen real estate. Furthermore, trying to tap a tiny dropdown arrow while walking to a meeting is an exercise in pure frustration.

The Solution:

1. **Use the Filter Pane:** Power BI Mobile has a built-in filter icon in the footer. Train your users to use it. It keeps your canvas clean and the data front-and-center.

2. **Horizontal Orientation:** If you must use on-screen slicers (like a "Year" or "Region" selector), change the slicer settings. Go to **Format -> Grid Layout** and switch from Vertical to **Horizontal**. This turns text lists into tappable buttons—much easier for a thumb to hit.

The Hardware Reality: Tablets vs. Phones

Finally, a point of confusion that trips up even experienced developers: **Tablets are not Phones.**

When you publish your report, the Power BI App behaves differently based on the hardware:

- **On a Phone (iPhone/Pixel):** The App looks for the **Mobile (Portrait)** layout you just built.
- **On a Tablet (iPad/Galaxy Tab):** The App treats the device as a small computer. By default, it renders the **Desktop (Landscape)** layout.

The Takeaway: Don't waste time trying to build a specific layout for an iPad. The iPad handles your desktop view just fine. Your energy must be focused entirely on the phone—the device in the pocket—because that is where the most critical, time-sensitive decisions are often made.

6.5 Exporting Options: PDF, PowerPoint, and Excel Live Connection

Let's be honest about your muscle memory. For years, "finishing" a report meant one thing: saving a static file, attaching it to an email, and hitting send. In that exact second, your data died. It became a snapshot, instantly disconnecting from reality while the business kept moving forward.

You need to reprogram that instinct. In Power BI, "exporting" isn't just about creating a copy. It is a strategic choice between **static artifacts** (for the record) and **dynamic windows** (for the conversation). Choose the wrong one, and you are back to manual version control hell.

1. PDF Export: The Static Snapshot

There are rare moments when data *must* stop moving. Audits. Month-end sign-offs. Regulatory filings. You need a "frozen moment" that no one can alter. This is the only scenario where a PDF is superior to the live report.

To get there, you navigate to **Export > PDF**. But pay attention to the nuance.

- **Current Values:** This captures your specific view. If you filtered for "Q3 Marketing Spend," that is exactly what the PDF locks in.
- **Default Values:** This hits the reset button. It strips away your analysis and reverts to the author's original view. Be careful. Do not send a generic report when you meant to send a specific insight.

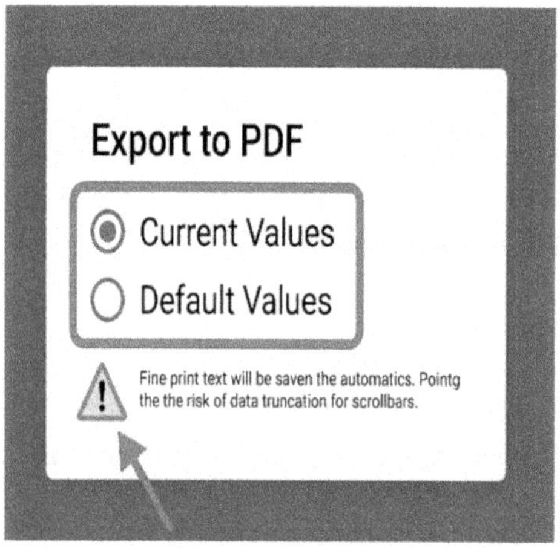

6.2 - ***The PDF Truncation Trap.*** *When exporting, always choose Current Values to capture the filters you have applied. Crucially, heed the warning (highlighted): PDF is a static medium. If you have a matrix table with a scrollbar, the PDF will only print the visible rows. The rest of your data will be cut off.*

The "Iceberg" Trap

Here is the limitation that burns most new users. It's painful. The PDF export is essentially a sophisticated screenshot tool. It captures the *visible* canvas.

If you have a table with 1,000 rows but only 15 are visible on the screen? The PDF will show those 15. The scrollbar does not expand. The rest of the data is lost. If you need the full list for an audit, PDF is useless. You need Paginated Reports or Excel.

2. PowerPoint: The End of the "Snipping Tool" Marathon

We have all been there. It is 9:00 PM before the Monthly Business Review. You are frantically taking screenshots of charts and pasting them into slides. By the time the meeting starts the next morning, the numbers are already stale.

Stop doing this. The **Power BI Add-in for PowerPoint** is your escape route.

You have two paths here.

- **The Legacy Image (Avoid this):** You can export pages as high-res images. It's fine for archiving, but in a meeting, it fails the "So what?" test.
- **Embed Live Data (The Game Changer):** instead of a dead image, you paste the report URL directly into the slide using the Add-in. The chart renders *inside* PowerPoint.

Why does this matter?

Imagine a VP asks, "Why is the variance so high in the North region?"

In the old world, you would apologize and promise to "get back to them." In the new world, you click the slicer *on the slide*. The charts update. You drill down. You answer the question in real-time. You look like a master of your data, not just a reporter of it.

The Socratic Truth regarding "Live" Data

Interactive presentations come with risks.

First, **Internet dependency**. If the Wi-Fi dies, your presentation goes blank. Second, **Licensing**. A pasted screenshot is visible to anyone. A live slide checks the viewer's credentials. If you email the deck to a CEO who lacks a Power BI license, they will see a "Permission Denied" error, not a revenue chart.

The Fix: Use the **"Snapshot"** feature in the Add-in. It freezes the view as an image for distribution but keeps the live connection available for your presentation. Best of both worlds.

3. Analyze in Excel: The Bridge to Your Comfort Zone

For the Finance Controller or Operations Analyst, this is arguably the most powerful feature in the entire ecosystem. It allows you to leave the Power BI interface but keep the Power BI brain.

The Amateur Move: "Export Data"

You hover over a visual, click the three dots, and select *Export data*. You get a CSV.

Do not do this.

This creates **Zombie Data**. It creates a file that is dead on arrival. It has no formatting, it is capped at 150,000 rows, and if the numbers change in the cloud five minutes later, your file is wrong. You are immediately back to managing "Version_Final_v2.xlsx".

The Pro Move: "Analyze in Excel"

This works differently. It downloads a tiny **.ODC (Office Data Connection)** file. When you open it, you see a blank Excel PivotTable.

The magic happens here. That PivotTable is connected via a secure tunnel directly to the Power BI dataset in the cloud. You drag "Sales" to Values and "Region" to Rows. The numbers appear instantly. But Excel isn't calculating them. Power BI is calculating them and streaming the result to you.

You get the flexibility of Excel—custom formatting, ad-hoc calculations, that familiar grid feel—with the governance of Power BI. When you refresh the PivotTable next month, the new numbers flow in automatically. No copy-paste required.

The "Grey Button" Frustration (And How to Fix It)

There is a specific, sinking feeling that happens when you follow a tutorial perfectly, only to find the critical button disabled. If you try to use "Analyze in Excel" and find it greyed out, take a deep breath: it is not a bug, and your software isn't broken. It is a strict governance setting.

To bridge Excel and the Power BI Cloud, you need what Microsoft calls **Build Permission** on the underlying dataset (the Semantic Model).

This is a significantly higher tier of trust than standard "Read" access. Here is the logic behind the lock: "Read" access simply allows a user to look at the pre-calculated, aggregated charts on a screen. "Build" access, however, gives you the keys to the engine room.

It allows your local Excel file to query the raw data and DAX measures directly. Because this allows users to potentially extract thousands of rows of granular data, IT departments and Workspace Admins usually disable it by default to protect data privacy.

*6.3 - **Analyze in Excel.** You don't have to force your finance team to give up Excel. By connecting a live PivotTable directly to the Power BI Cloud Service, they get the best of both worlds. Notice the highlighted Fields pane: those are not standard Excel columns. Those are the live tables (cylinders) and DAX Measures (calculators) streaming directly from your single source of truth in the Cloud.*

Day 7 - Smart Efficiency: Tools for the Power User

7.1 Using Parameters and Templates to Scale Reports

The Archimedes Principle of Data

Give me a lever long enough and a fulcrum on which to place it, and I shall move the world. In the context of data analytics, **Power BI is that lever.**

In the traditional spreadsheet world, effort is linear. You work for one hour; you produce one hour of output. If you need to produce the same report for ten different departments, you do the work ten times. That is not analysis; that is manual labor disguised as office work.

In Power BI, the math changes. You invest ten hours upfront to build a robust system, and that system generates thousands of hours of output over the next year. This shift—from *doing* the report to *designing* the engine that builds the report—is what distinguishes a junior analyst from a Senior Architect.

Today, we stop building one-off dashboards. We start building scalable data products. We will explore how to parameterize your work so a single file can serve infinite audiences, how to implement security so data protects itself, and how to use AI to find insights you didn't even know to look for.

The "Save As" Trap vs. The Scalable Solution

The Friday Afternoon Nightmare

It's 4:30 PM on the last Friday of the quarter. You've just finished the masterpiece: a "North Region" sales dashboard. The margins are calculated perfectly, the branding is spotless, and the conditional formatting highlights exactly what the VP needs to see.

Then, the email hits your Inbox: *"This is brilliant. Can you send me a version for the other 19 regional managers by 5:00 PM?"*

Most people panic. They walk straight into the **"Save As" Trap**.

They filter for "South," save as `Report_South.pbix`. They filter for "East," save as `Report_East.pbix`. Forty minutes of frantic clicking later, they have 20 individual files. They feel productive. But they aren't. They have just built a maintenance cage.

Two days later, you realize there was a typo in the Net Profit calculation. Now, you don't have one problem; **you have 20 problems.** You must open 20 files, make the same fix 20 times, and export 20 PDFs. This is linear effort. It is the absolute enemy of scalability.

The Solution: Decoupling Logic from Context

To escape the trap, you need to treat your report like an algebraic equation, not a static painting.

1. **Query Parameters (The Variables):** Think of your data source logic like the equation $Y = 2X + 5$. In a static report, you hardcode X as "North." If you want "South," you have to rewrite the equation. With Query Parameters, you leave X as a variable. You tell Power BI: "Load the data where Region equals $Parameter$." When you change the parameter, the entire data flow adapts instantly.

2. **Power BI Templates (The Blueprint):** A standard `.pbix` file contains both your report definition (charts, measures) and the actual data (which can be gigabytes). A Template file (`.pbit`) saves *only* the definition. It is a lightweight blueprint. When a user opens a template, it asks them: "Which Region do you want?" They type "West," and Power BI goes out and fetches *only* the West data.

Pro-Tip: The "Dynamic Path" Strategy

The number one error analysts face when sharing files is the **hardcoded file path**. If your Power BI file points to `C:\Users\YourName\Documents\Data.xlsx`, it will break the moment your colleague opens it on their machine.

Don't hardcode the path. Create a Text Parameter named `FilePath`. Use this parameter in your Power Query source step. When you send the template to a colleague, the very first thing Power BI will ask them is, "Where is the file located on *your* computer?" They paste their path, and the report updates without a single line of code breaking. This is how you build tools that other people can actually use.

1. The Engine: Query Parameters (Power Query)

Forget the code for a second. Think of a Parameter as a "Master Control Cell" in Excel. If you have a thousand formulas referencing cell `A1`, changing that single cell updates the entire spreadsheet instantly.

In Power BI, Parameters operate at the *connection level*, long before the data ever hits your charts. They are variables that sit inside Power Query and dictate *what* the system fetches.

Why this changes the game:

When you hardcode a file path (e.g., `C:\Users\YourName\Sales.xlsx`), you are essentially welding the report to your specific laptop. Send that file to a colleague, and it crashes instantly because their name isn't "YourName." By swapping that hard path for a Parameter, you turn a rigid link into a flexible variable.

The Strategy: Pre-Load Filtering

This is the hidden superpower of parameters that most beginners miss.

Instead of loading 10 million rows into Power BI and *then* filtering for the year 2024—which eats up your RAM and forces the user to wait—you pass the parameter `Year = 2024` to the database *before* the load begins. The database does the heavy lifting. Power BI only receives the specific slice you asked for.

It is the difference between trying to drink from a firehose and simply filling a glass from the tap.

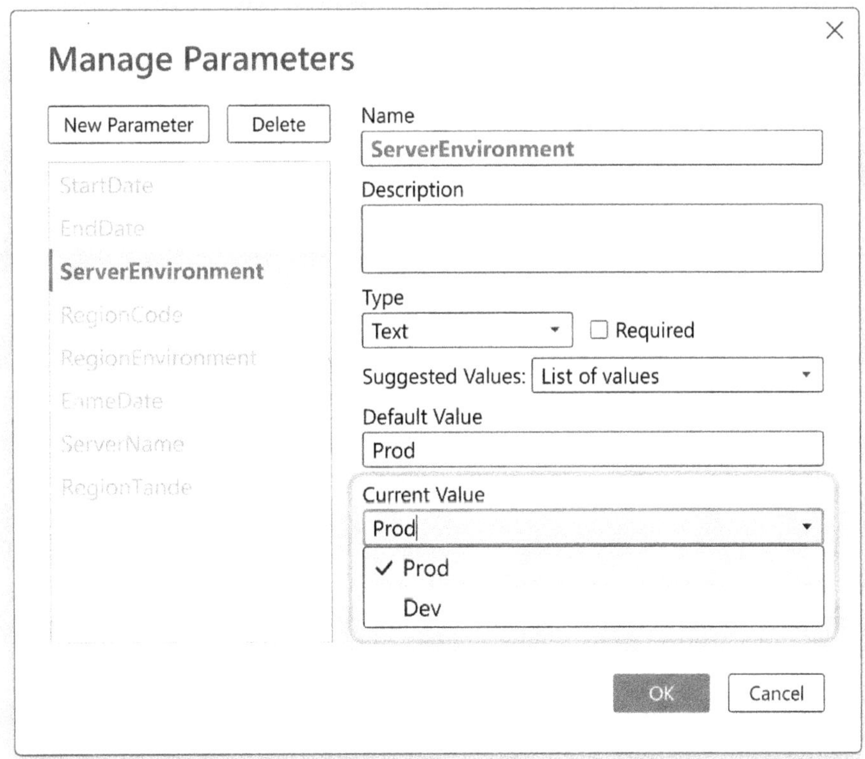

*7.1 - **The Parameter Switch.** Never hard-code a file path or server name directly into your queries. By replacing static text with a Parameter (like ServerEnvironment), you create a simple dropdown menu. Want to switch your entire report from the "Dev" testing database to the live "Prod" database? Change this single dropdown, click apply, and the engine repoints dozens of queries instantly.*

2. The Container: Power BI Templates (.pbit)

If a standard `.pbix` file is a fully furnished house crowded with people (your data rows), a **Power BI Template (.pbit)** is the architectural blueprint.

The Efficiency:

A standard `.pbix` file can be massive—hundreds of megabytes—because it stores the imported data inside itself. A `.pbit` file? It's often just a few kilobytes. It contains every visualization, every DAX measure, every color hex code, and every table relationship, but **zero data**.

The Interaction:

The magic happens the moment a user opens a template. Because the file is empty, Power BI knows it needs instructions. If you have set up Parameters correctly, the very first thing the user sees is not a blank screen, but a pop-up dialog box demanding input.

- "Which Region?"

DAY 7

- "Which Year?"
- "Which Server?"

The user types "West." Only then does Power BI reach out to the database, grab the data specifically for the West region, and populate the report.

The Reality Check: What They Don't Tell You

I'm not here to sell you a fantasy. While this architecture allows you to scale, there are three brutal truths you need to know before deployment:

1. **The "Credentials" Trap:** A template is not a bypass key. If you email a `.pbit` file connected to the company SQL database to a client who *doesn't* have a username and password for that database, the report will fail. Templates streamline structure, they do not circumvent security.

2. **Case Sensitivity Kills:** Computers are literal to a fault. If your parameter filters for "North" (capital N), and the user types "north" (lowercase n), they might get a blank report. Power Query is case-sensitive. You must account for this in your logic (perhaps by forcing inputs to uppercase) or your users will assume the report is broken.

3. **The "One-Way" Street:** Once a user opens a `.pbit` and loads the data, it transforms into a standard `.pbix` file. They can save it. They can edit it. But if you update the Master Template next week, their saved file does *not* automatically update. You are distributing copies, not creating a live link to the master design.

3. The Scalable Workflow

Despite the caveats, this architecture is the only way to scale yourself from a "Report Builder" to an "Analytics Architect."

Here is the workflow for the efficient Power User:

1. **Define:** You create the master report. You set a Parameter in Power Query called `RegionName`. You filter your main data table using this parameter.

2. **Export:** You do not "Save As." You go to `File > Export > Power BI Template`.

3. **Distribute:** You email this tiny 50KB file to 20 managers. It doesn't clog inboxes; it barely takes up space.

4. **Instantiate:** The "North" manager opens the file. Power BI prompts: "Enter Region." They type "North."

5. **Result:** Power BI queries the database specifically for North data. The manager saves their file as `North_Q3.pbix`.

You have just serviced 20 stakeholders with a single asset, zero version conflicts, and minimal maintenance.

7.2 Row-Level Security (RLS): Controlling Who Sees What

The End of the "Save As..." Nightmare

For decades, Excel users have managed data security through a process of manual fragmentation. If you have sales data for five regions, you likely create five separate files: `Sales_North.xlsx`, `Sales_South.xlsx`, and so on. You password-protect them, email them out, and pray you didn't accidentally paste the wrong rows into the wrong file.

This method is the "physical key" approach. It is brittle, time-consuming, and prone to human error. If a manager moves from North to South, you have to physically retrieve the old file and issue a new one.

Row-Level Security (RLS) eliminates this entirely. In Power BI, you maintain **one single dataset** and **one single report**. You do not split files. Instead, you change how the user interacts with that file.

The Hotel Key Card Principle

To understand RLS, stop thinking like an Excel user and start thinking like a Hotel Security Manager.

The Excel Method (Physical Keys):

Imagine a hotel with 50 rooms. In the old world, you give every guest a specific metal key that opens only Room 101. If the guest moves to Room 102, you must take back the first key and cut a new one. If you renovate the hotel (update the report logic), you have to go to every room to make sure the keys still work.

The Power BI Method (The Master System):

RLS is a modern magnetic key card system. Every user holds the exact same plastic card (the Report). When they swipe it at the elevator, the central system checks their ID.

- User A swipes: The system recognizes them as "North Manager" and the elevator only allows them to press the button for the 2nd floor.
- User B swipes: The system sees "Global VP" and unlocks the Penthouse (all data).

The building (the data) is the same for everyone. The card (the report) is the same. The only thing that changes is the invisible permission rule in the central server.

The Two Architectures: Hard-Coding vs. Scalability

When you sit down to design security, you are at a crossroads. One path is quick to build but excruciating to maintain. The other requires a Data Architect's mindset but scales infinitely.

A. Static RLS (The Manual Method)

This is the intuitive approach. It's also usually a trap. You hard-code the logic directly into the report.

DAY 7

- **The Logic:** You create a specific Role for each segment (e.g., "North Manager"). The DAX filter is simple: ` [Region] = "North" ` .
- **The Reality:** This works beautifully if you have a stable team of five people covering four fixed regions.
- **The Brutal Truth:** In a growing organization, Static RLS is a maintenance nightmare. If you hire a new manager for a "Midwest" territory, you cannot simply update a user list. You must open the Power BI Desktop file, create a new "Midwest" role, write the DAX, save, and republish the entire .pbix file. If you have 50 branch managers, you are effectively creating 50 separate maintenance tasks for yourself.

B. Dynamic RLS (The Scalable Method)

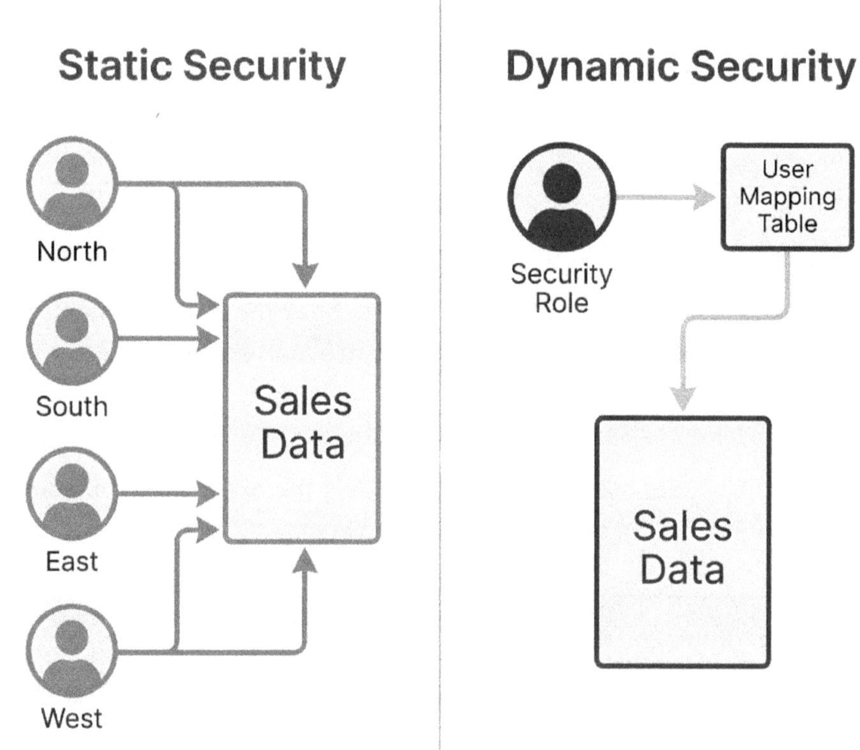

*7.2- **Hard-Coding vs. Scalability.** Static RLS (Left) is like cutting a physical key for every single door: you must create and manage a new role for North, South, East, and West. Dynamic RLS (Right) uses the "Hotel Key Card" principle. You build only one Security Role. That role checks the User Mapping Table to see who is logging in (e.g., USERPRINCIPALNAME()), dynamically filtering the data downstream.*

This is how you should be thinking. Instead of hard-coding "North," you tell Power BI: *"Find out who is logging in, look up what they are allowed to see in a reference table, and filter the data accordingly."*

- **The Mechanism:** You create only *one* role (e.g., "Security_Role").
- **The Golden Function:** ` USERPRINCIPALNAME() ` .

- *Critique:* You will be tempted to use `USERNAME()`. Don't. In Power BI Desktop, `USERNAME()` returns your local computer format (`DOMAIN\User`), but in the Service, it returns your email (`User@Company.com`). This mismatch breaks your logic the moment you publish. `USERPRINCIPALNAME()` returns the email address in both environments. It is the only reliable standard.
- **The Logic:** Your DAX filter on a hidden User Table looks like this: `[Email_Address] = USERPRINCIPALNAME()`.

The Implementation Checklist

Implementing Dynamic RLS requires you to wire your data model differently. Since you are likely filtering a large Fact Table based on a tiny User Table (which usually filters a Dimension Table), the signal needs to travel correctly.

1. **The Directionality Trap:** By default, relationships flow "One-to-Many." However, security filters often need to flow from a User Table *up* to a Dimension and then *down* into the Fact Table.

    - *The Fix:* When editing the relationship between your User Table and your Data/Dimension table, you must check the box **"Apply security filter in both directions."** If you miss this checkmark, the user will log in, the User Table will filter successfully, but the Sales Data will remain completely visible. The bridge is out.

2. **Test Locally (The Flight Simulator):** Never publish blindly. In Power BI Desktop, go to `Modeling > View As`.

    - If using Static RLS, check the specific Role.
    - If using Dynamic RLS, check the Role *and* check "Other user," then manually type the email address of a colleague.
    - If the report doesn't filter exactly as expected here, it will fail in the cloud.

The Deployment "Gotcha": The Workspace Paradox

You have written the perfect DAX. You have tested the roles. You publish the report. Then, you get an angry email: *"I set RLS for the Sales Manager, but she can still see the CEO's data."*

Here is the truth that the documentation often buries: **Workspace Permissions override Row-Level Security.**

If a user has "Edit" rights, Power BI assumes they are a developer. Developers need to see all data to verify calculations. Therefore, RLS is *ignored* for anyone with the following Workspace roles:

- **Admin**
- **Member**
- **Contributor**

RLS *only* applies to users with the **Viewer** role in the Workspace, or users consuming the report via an **App**.

The Strategic Advice: Do not use the Workspace as a shared folder for end-users. The Workspace is your construction site; only give hard hats (Admin/Member/Contributor access) to your fellow builders. For everyone else—the consumers—publish an **App** and grant them access there. The App enforces the "Viewer" status, ensuring your invisible filters actually work.

	Admin	Member	Contributor	Viewer
Can Edit Report	Yes	No	No	Yer
Can Publish	Yes	Yes	Yes	No
RLS Applied?	NO	NO	NO	YES

*7.3 - **The RLS Bypass Trap.** Do not give end-users the "Contributor" role just to look at a report. As this matrix shows, any workspace role with editing privileges (Admin, Member, Contributor) automatically ignores Row-Level Security (RLS). To enforce data restrictions, your audience must be assigned the Viewer role or access the report via a published App.*

7.3 AI Visuals: Key Influencers and Decomposition Trees

The Concept: From Mirror to Detective

Let's be honest. Most standard visualizations are passive. They are mirrors. You feed them data, and they reflect it back to you exactly as it is. If sales are down, a bar chart shows you a shorter bar. It doesn't tell you *why*. In the Excel world, finding that "why" is an exercise in brute force.

You see a variance. You create a Pivot Table. You drag "Region" to the rows. Nothing obvious. You swap it for "Product." Then "Salesperson." Then "Month." It is a manual, tedious game of "Go Fish." You are guessing, hoping to stumble upon the driver of the variance.

AI Visuals invert this dynamic.

Instead of you telling the visual what to display, you tell the visual what result you are interested in—for example, "Why is Profit down?"—and it scans your *entire* dataset behind the scenes. It runs thousands of permutations in milliseconds to tell you which dimensions matter. It replaces the "hunch" with statistical probability.

The Key Influencers Visual: Automated Driver Analysis

This visual is essentially a **"Data Scientist in a box."**

It democratizes complex regression analysis. You might be accustomed to looking for correlations by eye, but human eyes are biased; we see patterns where there are none. The Key Influencers visual uses **ML.NET** (Machine Learning for .NET) to run logistic regression (for binary categories like Churn/No-Churn) or linear regression (for continuous numbers like Sales).

It mathematically ranks factors by how much they push your metric up or down.

The Experience

Imagine looking at a dashboard showing a spike in customer churn. Instead of guessing, you look at the Key Influencers bubble. It tells you explicitly:

"When 'Contract Type' is 'Month-to-Month', the likelihood of Churn increases by 2.4x."

It strips away the noise and hands you the signal. It turns a wall of data into a ranked list of suspects.

How to Set It Up

- **Analyze:** This is your target. Drag the metric that keeps you up at night here (e.g., `[NPS Score]` or `[Late Shipment]`).
- **Explain By:** This is where you dump the haystack to find the needle. Drag every dimension you suspect might be involved (e.g., `[Country]`, `[Vendor]`, `[Day of Week]`, `[Discount %]`). The engine will sort them out for you.

*7.4 - **Feeding the AI.** The Key Influencers visual requires no code. You simply tell the engine what metric you care about in the Analyze bucket (e.g., Total Sales), and throw every possible dimension you can think of into the Explain By bucket. The AI runs a machine-learning regression model in seconds, isolating the exact factors that drive your sales up or down.*

The Brutal Truth: Where It Fails

Here is what the sales brochures won't tell you. **This visual is dangerous if you turn your brain off.**

1. **The "Ice Cream & Sharks" Problem:** The AI finds mathematical correlation, not real-world causation. It might tell you that "High Ice Cream Sales" causes "Shark Attacks" because both numbers go up in July. It is up to *you*, the human expert, to know that the real driver is "Summer." Do not blindly report the AI's findings without applying business logic.

2. **The "Small Data" Trap:** This is not for your small, 50-row pilot project. Statistics require volume to be significant. Microsoft generally recommends at least **100 observations** for the outcome you are analyzing. If you filter your report down to a specific niche with only 15 rows of data, the visual will either throw an error or, worse, give you a misleadingly confident answer based on insufficient evidence.

3. **Garbage In, Garbage Out:** If your data isn't clean—for example, if you have "USA," "U.S.A.", and "United States" as three different values—the AI will split the influence across them, diluting the insight. You cannot skip the data modeling phase.

The Decomposition Tree: AI-Powered Root Cause Analysis

Standard drill-downs are rigid. You build a hierarchy: *Year > Quarter > Month*. But what if the anomaly isn't in a specific month, but in a specific *Product Category*? The Decomposition Tree is a **"fluid hierarchy."** It allows you (or your end-user) to decide the path of exploration in real-time. It visualizes the contribution of individual parts to the whole, maintaining the context of the total sum as you branch out.

The Experience

Think of this as a "Choose Your Own Adventure" book for your data. You start with `Total Profit`. You see a small `+` sign. When you click it, you aren't forced to go to `Region`. You can choose `Salesperson`, `Product`, or `Store Type`. As you expand the branches, you create a visual map of where the money is going. It is tactile, interactive, and incredibly satisfying for presentations because it allows you to answer executive questions on the fly.

The AI Split (The Lightbulb)

This is the "Show, Don't Tell" feature. When you click the `+` to drill down, you will see options marked with a lightbulb: **"High Value"** and **"Low Value."**

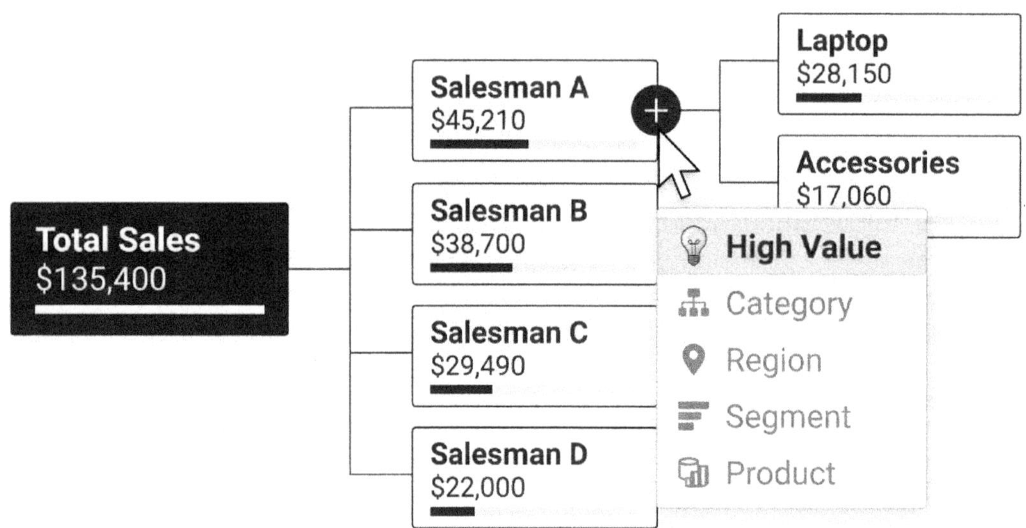

7.5 - ***The AI Split.*** *Instead of manually guessing which dimension (Region, Product, etc.) to drill into next, click the + icon and look for the lightbulb. Choosing High Value or Low Value instructs the machine learning algorithm to instantly scan all your data and automatically expand the category that has the single biggest impact on your Total Sales.*

If you select **High Value**, the AI scans all the dimensions you made available and automatically picks the one that causes the biggest spike.

It's like a flashlight pointing immediately to the biggest contributor to your problem, saving you from clicking through ten different dimensions to find the variance.

The Brutal Truth: Where It Fails

While visually stunning, this tool has "gotchas" that frustrate professionals:

1. **Screen Real Estate:** A tree can grow very wide, very fast. On a standard laptop screen, a four-level tree becomes unreadable. It is fantastic for deep-dive analysis, but terrible for a high-level executive summary dashboard where space is at a premium.

2. **The 5000 Point Limit:** The visual has a rendering limit of 5,000 data points. If you try to decompose a tree by "Transaction ID" with millions of rows, the tree will truncate. You might miss the outliers simply because the visual stopped rendering.

3. **Export Frustration:** If you rely on "Publish to Web" (public links) or an on-premise Report Server, the fancy "AI Split" (High/Low value) functionality is often disabled or restricted. Check your distribution method before falling in love with the feature.

Summary: Which Tool When?

- Use **Key Influencers** when you want the answer handed to you. It is for **ranking hidden drivers** (e.g., "Rank the top 5 reasons our NPS score dropped").
- Use the **Decomposition Tree** when you want to explore the journey. It is for **ad-hoc root cause analysis** (e.g., "Let me drill down into the West region to see which specific store is bleeding money").

7.4 Performance Analyzer: Why Is My Report Slow?

There is a specific, hollow silence that fills a boardroom when a report fails to load. Everyone is staring at your screen. You are staring at a spinning wheel. The silence stretches. In that moment, the instinct is self-preservation. You blame the Wi-Fi. You blame the VPN. You blame the server capacity or the sheer size of the dataset.

Here is the brutal truth: 90% of the time, the slowness isn't coming from the IT infrastructure. It is coming from the report itself.

Power BI is a high-performance engine. But like a Ferrari, if you put square wheels on it, it won't move. To fix the speed, you have to stop guessing. You need the flight recorder. You need the **Performance Analyzer**.

DAY 7

1. The Flight Recorder (Stop Guessing)

In Excel, when a file hangs, you are helpless. You sit there, watching the "Calculating: 40%" bar creep forward, hoping it doesn't crash. Power BI is different. It includes a built-in diagnostic tool that acts exactly like a black box on an airplane. It records every millisecond the engine spends thinking, rendering, or waiting.

Why is this necessary? Because a report page isn't one single image. **It is a collection of independent queries.**

If you have 10 charts on a page, Power BI is effectively running 10 separate questions against the database simultaneously. It's a race. The Performance Analyzer breaks down that race, telling you exactly which runner is limping. To access it, go to the **View** tab and check the **Performance Analyzer** box. A new pane will dock on the right side of your canvas. It looks deceptively simple. It isn't.

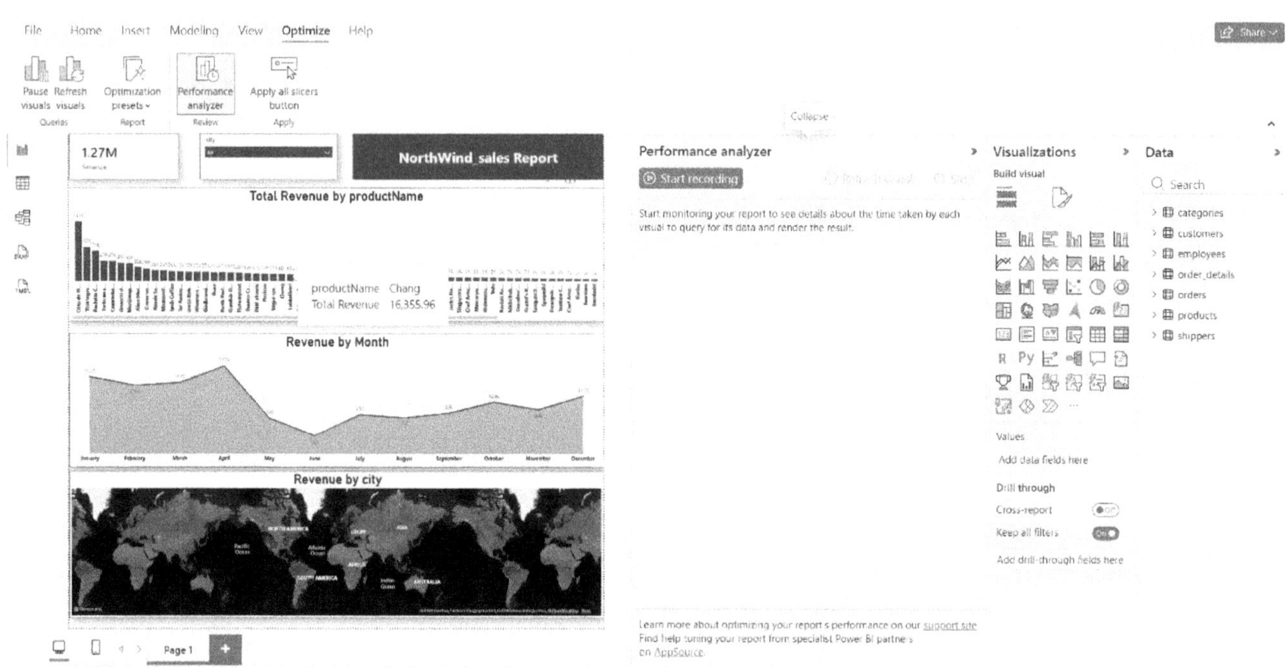

*7.6 - **Stop Guessing, Start Measuring.** If your report is slow, do not guess why. Open the View tab and launch the Performance Analyzer. Clicking "Start recording" turns Power BI into a diagnostic engine, tracking exactly how many milliseconds it takes to render every single visual on your page.*

2. The Testing Protocol (The "Clean Bench" Rule)

Most users open the pane and just stare at it. This is useless. To get a scientific reading, you must follow a strict sequence. If you don't, you are measuring noise, not signal.

1. **Open the Pane:** Ensure the Performance Analyzer is visible.
2. **Start Recording:** Click the distinct **Start recording** button.
3. **The Critical Distinction:** Click **"Refresh visuals"** *inside the analyzer pane.*

Read that last point again. Do **NOT** click the "Refresh" button on the Home ribbon.

The button on the Home ribbon triggers a **Data Refresh** (reaching out to your SQL server or Excel file to check for new rows). That tests your internet speed. The button in the pane triggers a **Visual Refresh** (re-calculating the math on the data you already have loaded). We are testing your DAX, not your Wi-Fi. Watch the milliseconds tick up. Once finished, click **Stop**.

3. Decoding the Metrics (The Three Buckets of Time)

Expand any visual in the list (e.g., "Sales by Month"). You will see a stacked bar chart with three distinct colors. This is your Root Cause Analysis.

A. DAX Query (The Calculation)

This represents the time the engine spends "doing the math." It is scanning columns, applying filters, and summing rows.

- **The Diagnosis:** If this bar is long (Red), your logic is too complex. You might be asking Power BI to iterate through a million rows to find a single value.
- **The Fix:** This is a formula problem. You need to optimize your DAX measure.

B. Visual Display (The Rendering)

This is the time spent "painting" the pixels on the screen *after* the math is finished.

- **The Diagnosis:** If this bar is long (Green), you are overloading the browser. A common culprit is a Map with 30,000 unaggregated pinpoints, or a Scatter Plot with too many dots. The math was fast, but the human eye—and the computer's graphics card—can't process that much density.
- **The Fix:** Summarize the data. Don't show every transaction; show the trends.

C. Other (The Queue)

This is the most confusing and frustrating metric. **"Other" essentially means Waiting in Line.** Here is the truth nobody tells you: Power BI does not load everything at once. It creates a queue. It typically processes about 6 visuals in parallel. If you have a dashboard with 40 "Card" visuals (KPIs), the 40th card sits in the "Other" bucket doing absolutely nothing while it waits for the first 6 to finish. If "Other" is high, you have too many visuals on the page. Consolidate. Merge your 10 separate Card visuals into one "Multi-row Card" or a Matrix.

4. The "Cache" Trap (Don't Be Fooled)

Power BI is designed to be smarter than us. If you ask it to calculate "Total Sales" and it takes 2 seconds, it stores that answer in a short-term memory bank called the **Cache**. If you change a slicer and then change it back immediately, the report will load instantly (0 ms).

The Trap: You might think, "Great, my report is fast now!" No, it isn't. The engine just recited the answer from memory. It didn't do the work. This is why using the **"Refresh visuals"** button in the pane is non-negotiable. It forces the engine to dump its memory and calculate the result from scratch, giving you the honest "Cold Cache" performance time.

5. The Bridge to Advanced Engineering

Sometimes, you will see a massive **DAX Query** bar, and you won't know why. You typed `SUM(Sales)`, so why is it taking 4 seconds? Power BI gives you a backdoor to the code. Hover over the slow visual in the list and click **"Copy query"**.

This copies the raw SQL-like command the visual is sending to the engine. While you might not know how to read this yet, this text is the key to the kingdom. You can paste this into external tools like **DAX Studio** (a tool for the true engineers) to perform surgery on your code line-by-line. But for now, just knowing *which* visual is the bottleneck puts you miles ahead of the average user who simply blames the server.

7.5 Bookmarks and Buttons: Creating App-Like Navigation

Stop building reports that act like PowerPoint slides. Seriously. If your users are navigating your report by clicking those tiny, clumsy tabs at the bottom of the screen, you are leaving engagement on the table. Those tabs are passive. They are rigid. And let's be honest, they are boring. We need to shift your mindset from building "pages" to building **applications**. You want a report that behaves like modern software. We are talking about clickable menus, help overlays that pop up on command, and toggle switches that swap complexity for simplicity. To do this, we rely on the central nervous system of Power BI interactivity: **Bookmarks** (the memory bank) and **Buttons** (the triggers).

The "Easy Button": Navigators

Before you open Photoshop to design custom icons for every single page, stop. The "old way" of navigation involved creating individual buttons, manually linking them, and updating every single page whenever you added a new tab. It was a maintenance nightmare. A total waste of time. Microsoft eventually gave us a lifeline: **Navigators**.

The Page Navigator

This tool replaces those bottom tabs with a web-style navigation bar.

- **The Logic:** It reads your report's schema dynamically. It generates a button for every visible page automatically.
- **The Benefit:** If you rename "Sales" to "Revenue" or delete a tab entirely, the buttons update instantly. No broken links. No manual grunt work.

The Bookmark Navigator

Same principle, different target. Think of an Excel file where you want to swap a chart from showing "Region" to "Product." You don't need two separate tabs. The Bookmark Navigator creates a sleek toggle bar that switches the view without the user ever leaving the screen.

The Core Concept: It's a "Save State"

If you take nothing else from this section, burn this into your brain: **A bookmark is NOT a hyperlink.** It is a "Save State." Think of it like a "Save Point" in a video game. When you click "Add Bookmark," Power BI takes a forensic snapshot of the page's current existence. It records three specific layers of reality:

1. **Visibility:** What is shown versus what is hidden.
2. **Data State:** Which filters, slicers, and cross-highlights are active.

3. **Sort Order:** How your tables are organized.

To control this, you must master the **Selection Pane**. If the Report View is the stage, the Selection Pane is the backstage area where you decide which props are behind the curtain and which are under the spotlight.

The "Data" Trap (Read This Carefully)

Here is the truth most tutorials gloss over: **Default bookmarks are dangerous.** By default, Power BI saves *everything*. It snapshots the fact that you were looking at "2023 Data" for "North America" when you created the bookmark. If you attach that bookmark to a navigation button, you are setting a trap. Imagine a user analyzing **2024** data. They click your "Help" button to see a definition. If that bookmark saved the Data state, the report will violently jerk the user back to the filters *you* selected—2023, North America. The user loses their place. They get frustrated. They blame the tool. To fix this, right-click every bookmark and be surgical about what it remembers:

- **Data (The Filter Trap):**
- **Uncheck this** for UI elements (pop-ups, toggles). You want the user's filters to persist.
- **Check this** ONLY for "Reset" buttons where the goal is to clear filters.
- **Display:**
- **Check this** for almost everything. This is the engine that remembers "Chart A is hidden, Chart B is shown."
- **Current Page:**
- **Check this** only if the button moves the user to a new screen.

Scope: The Nuke vs. The Scalpel

The second failure point is the scope of your snapshot.

- **All Visuals (The Nuke):** The bookmark records the state of every single element on the page. If you add a new KPI card later, this bookmark might mess with it unexpectedly.
- **Selected Visuals (The Scalpel):** This is the professional standard. You select only the specific elements you want to change (Ctrl+Click the visuals), then update the bookmark to "Selected Visuals."

Workflow 1: The "Reset All" Panic Button

This is the highest-ROI button you can build. When users get lost in a web of drill-downs and slicers, they need a way out.

1. **Set the State:** Manually clear every slicer. Set the report exactly how it should look on Day 1.

2. **Capture:** Create a bookmark named "Reset Filters".

3. **Configure:** Right-click the bookmark. Ensure **Data** is Checked (you want to overwrite their filters). Ensure **All Visuals** is Checked (you want to reset the whole page).

4. **Trigger:** Insert a standard "Reset" button and link it to the bookmark.

Workflow 2: The Visual Toggle (Chart vs. Table)

Knowledge workers love tables. Executives love charts. Instead of arguing, give them both in the same pixel space.

1. **Setup:** Stack a Chart and a Matrix (table) directly on top of each other. Open the **Selection Pane**.
2. **Snapshot A (The Chart View):**
    - Hide the Matrix (click the eye icon). Show the Chart.
    - *Crucial:* Select **both** the Chart and the Matrix in the Selection Pane (Ctrl+Click).
    - Add Bookmark: "View Chart".
    - Update settings: **Uncheck Data** (preserve filters). Select **Selected Visuals** (don't touch the rest of the page).
3. **Snapshot B (The Table View):**
    - Hide the Chart. Show the Matrix.
    - Select **both** visuals again.
    - Add Bookmark: "View Matrix".
    - Update settings: **Uncheck Data**. Select **Selected Visuals**.

Link these to a Navigator. You have now created an interface that responds to the user, respects their context, and doubles your screen real estate. You aren't just reporting data anymore; you're designing an experience.

Conclusion: Your Path Forward

Recap: The Transformation from Analyst to Data Storyteller

You didn't start this journey just to learn where the buttons are located in a new piece of software. You began this process because you were frustrated. You were a Knowledge Worker—a Controller, a Marketing Analyst, perhaps an Operations Manager—armed with high analytical intelligence but shackled by the tools at your disposal.

This isn't just a summary of technical features. It is a validation of your professional evolution. We are mapping your escape from the production-heavy "Report Factory" to the strategic role of "Data Storyteller."

Here is how your professional identity has shifted.

1. The "Report Factory" Trap (The Past)

Let's be honest about the "good old days." They weren't good. Before mastering the modern data stack, the typical day of a smart analyst was paradoxically spent doing "dumb" work.

Traditional Analyst	Data Storyteller
Focus on What	Focus on Why & Action
Static Reports	Interactive Dashboards
Manual Data Prep	Automated Pipelines
Email Silos	Single Source of Truth

*Z.1 - **The Career Pivot.** Power BI is just a tool; the real goal is transforming your professional identity. The Traditional Analyst (Left) is stuck in a low-leverage loop of manual mechanics, answering "What happened?" by sending static files. The Data Storyteller (Right) builds high-leverage assets. You automate the mechanics so you can focus on the "Why," delivering a single source of truth that drives actual business action.*

We call this the **80/20 Efficiency Paradox**. You burned 80% of your cognitive fuel on data *production*—extracting CSVs, cleaning messy columns, fixing broken VLOOKUP ranges, and battling formatting alignment—leaving a pitiful 20% for actual *analysis*.

The Diagnosis: This inefficiency stems from the "fragility of the grid." In Excel, data and presentation are fused in the same cells. Every time new data arrives, the structure breaks, forcing you into a reactive cycle of maintenance.

The Reality: Remember the anxiety of "Version Control Hell"? *Q3_Report_Final_v2_REAL_FINAL.xlsx*. The sinking feeling of clicking "Send" and praying you didn't miss a row in your sum range? You weren't delivering insights; you were delivering rows and columns. You were handing stakeholders a haystack and wishing them luck in finding the needle.

2. The Catalyst: Power BI as the "Time Dividend"

Here is the secret Microsoft won't put on a billboard: Power BI is not primarily a visualization tool. It is an **automation engine**. By utilizing Power Query and the Data Model (Star Schema), you have effectively separated the *definition* of the data from the *presentation* of the data.

- **The Science:** You replaced fragile cell references with robust relationships. You moved from static arithmetic to DAX—dynamic calculations that respond to user interaction.
- **The Brutal Truth:** Your CFO doesn't care about your Star Schema. They don't care how elegant your DAX code is. They care about *speed to insight.*

The only reason to master these technical skills is to generate a "Time Dividend." By automating that 80% of manual drudgery, you reclaim the hours necessary to do the hard work: thinking. Gartner predicts that by 2026, data storytelling will be the dominant means of consuming analytics.

Why? Because as data volume explodes, the ability to summarize it decreases. Automation is the only survival strategy.

3. The Destination: What is a Data Storyteller?

We need to kill a misconception right now. **Data Storytelling is not about making pretty charts.**

You can build a beautiful dashboard that says absolutely nothing. That isn't storytelling; that is "data decoration." It is noise. True Data Storytelling is a structured form of communication designed to bridge the gap between logic and emotion. Humans justify decisions with logic (Data), but we make decisions based on emotion and context (Narrative).

The Brent Dykes Framework:

To move from "Number Cruncher" to "Strategic Advisor," you must combine three elements:

1. **Data:** The foundation. It provides **Accuracy** and trust.

2. **Visuals:** The enlightenment. It allows the eye to see patterns, trends, and outliers that remain invisible in a spreadsheet.

3. **Narrative:** The engagement. This is the context. It explains the "So What?"

The Socratic Check: Ask yourself before publishing: "Does this report just show *what* happened, or does it guide the user toward *what to do next*?" If it lacks the narrative arc—the explanation of cause and effect—it is still just a report, no matter how interactive it is.

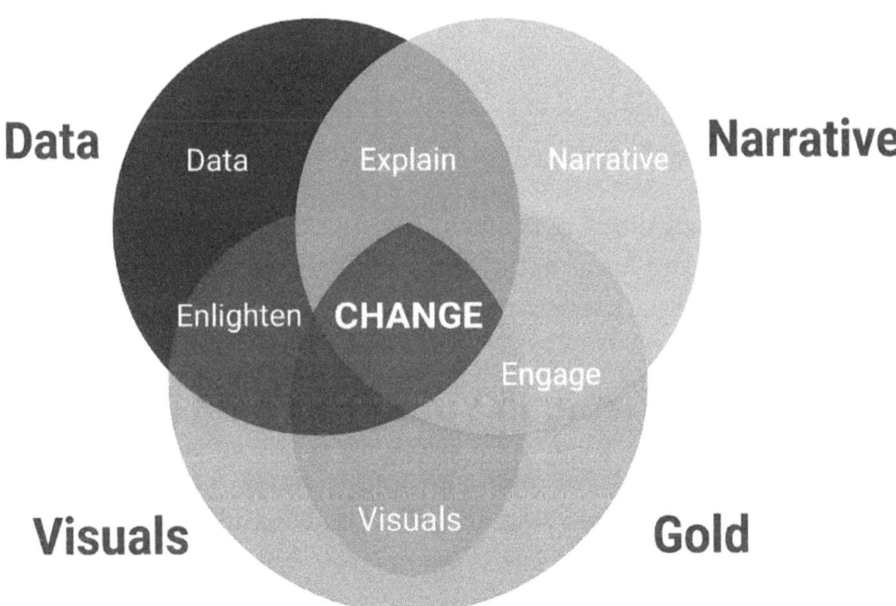

*Z.2 - **The Architecture of Impact.** Data alone is just noise. Visuals alone are just pretty pictures. A narrative without data is just an opinion. As a modern Data Storyteller, your job is to stand exactly in the center of this diagram. When you combine robust data models, clean visual engineering, and business context, you don't just report the news—you drive Change.*

4. The New Competency Stack: The "Full-Stack" Analyst

You are no longer just an Excel user. You are developing a "Full-Stack" competency that blends the technical with the psychological.

- **Data Engineering (Lite):** You use Power Query to structure reality into a format a machine can read.
- **Semantic Modeling:** You define the "Single Source of Truth," eliminating the arguments over whose numbers are correct.
- **Design Thinking:** You understand that a dashboard is a User Interface. You choose visuals not because they look cool, but because they reduce the time it takes for a brain to process information (Cognitive Load theory).
- **Communication:** You draft the narrative. You frame the problem.

The Future-Proofing Reality

Let's address the elephant in the room: **AI (Copilot).**

AI is getting exceptionally good at writing DAX and generating charts. If your value proposition is only "I can make a bar chart," you are at risk.

However, AI is terrible at context. It does not know office politics; it does not know that the sales dip in Q3 was due to a competitor's pricing change that wasn't captured in the SQL database.

Your competitive advantage is no longer the code; it is the context. Your ability to weave AI-generated insights into a coherent business narrative is the one skill that cannot be automated.

5. The Permanent Shift

The transformation is done. You have stopped sending email attachments (static artifacts of the past) and started publishing Apps (dynamic engines of truth).

The difference between a mediocre analyst and a great one is not their knowledge of functions. It is their ability to influence change. You have moved from the back office of "Reporting" to the front lines of "Decision Making." Do not go back to the factory.

Common Pitfalls to Avoid in Your First Real Project

You have the technical basics. You understand the syntax. But let's be clear: the transition from "Excel User" to "Power BI Analyst" is not about learning DAX. It is about unlearning habits. The behavioral traps waiting for you aren't syntax errors; they are structural mistakes born from years of spreadsheet-based thinking.

In Excel, you are the master of the cell. You control the grid. In Power BI, you must become the architect of a model. The very skills that made you efficient in spreadsheets—visualizing data as a wide grid, hardcoding quick fixes, and creating massive "master tables"—are the exact instincts that will cause your Power BI reports to grind to a halt.

Here are the four "Excel Instincts" you must actively suppress.

1. The "Flat File" Fallacy (The Excel Legacy)

The Why:

In Excel, performance is rarely impacted by the width of your table. You are used to seeing everything in one place: Customer Name next to Order Date next to Product Category. This creates a comforting sense of control.

But Power BI's engine, **VertiPaq**, hates this. It is a columnar database that compresses data by looking for unique values in columns.

When you import a single, massive, 50-column "flat" table, you force the engine to scan millions of duplicate text values. Imagine repeating "North Region" for every single transaction in a million-row dataset. This kills compression. It destroys calculation speed.

The Experience:

Imagine you are tasked with packing a library.

- **The Excel/Flat File way:** You write the full author biography and the publisher's address on the inside cover of *every single book*. If you have 1,000 books by Stephen King, you write his bio 1,000 times. It's heavy, redundant, and wasteful.
- **The Power BI way:** You write the bio once on a card index. Inside the book, you just put a reference number: "Author ID #5."

When you ask Power BI to "Count books by Author," the Flat File method forces it to read thousands of text strings. The Star Schema method simply counts the IDs. It is lightweight. It is fast.

The Truth:

Let's be honest about why you build flat files. You don't trust the relationships. You want to *see* the "Customer Name" right next to the "Sales Amount" in the Data View because it feels safe. You are afraid that if you split the data, you'll lose control. But by clinging to this safety blanket, you are creating a report that will become unusable the moment your data grows beyond a trivial size. If you have to scroll horizontally to see your data, you've built it wrong.

The Fix:

Split your data. Keep transactions in a **Fact Table** (Sales, quantities, dates) and descriptive attributes in **Dimension Tables** (Customers, Products, Locations).

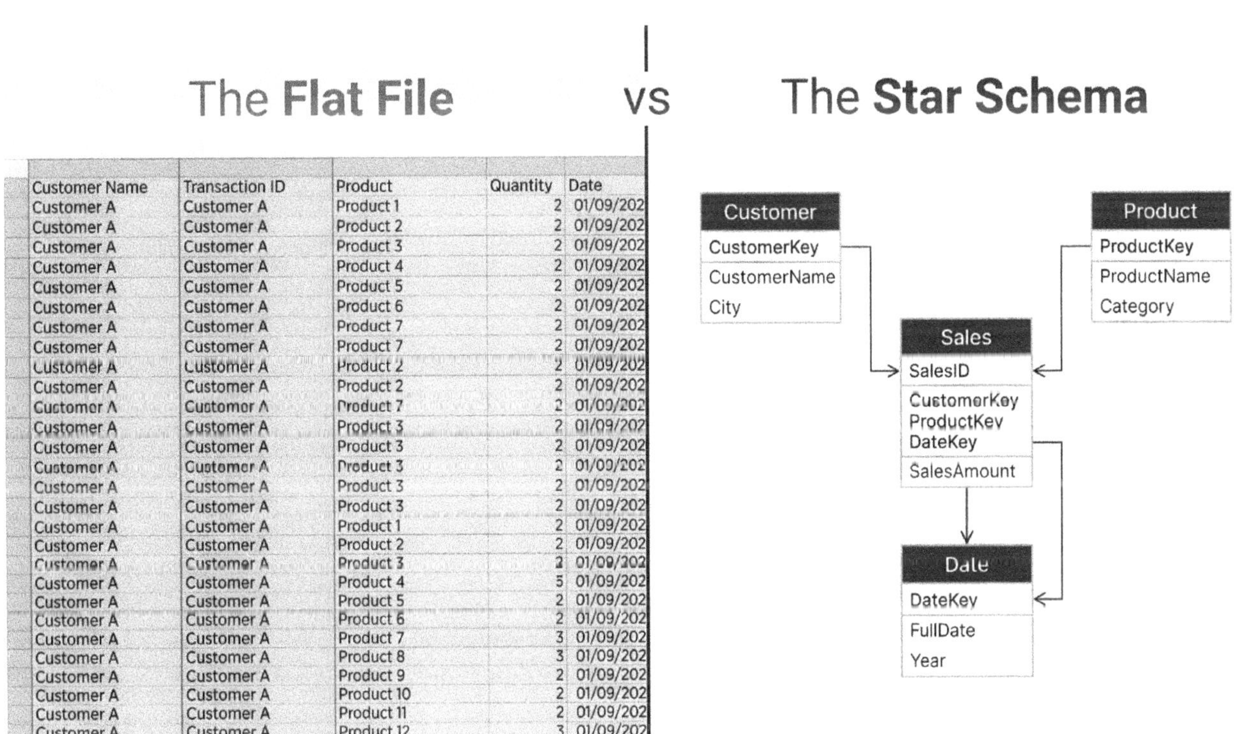

Customer Name	Transaction ID	Product	Quantity	Date
Customer A	Customer A	Product 1	2	01/09/202
Customer A	Customer A	Product 2	2	01/09/202
Customer A	Customer A	Product 3	2	01/09/202
Customer A	Customer A	Product 4	2	01/09/202
Customer A	Customer A	Product 5	2	01/09/202
Customer A	Customer A	Product 6	2	01/09/202
Customer A	Customer A	Product 7	2	01/09/202
Customer A	Customer A	Product 7	2	01/09/202
Customer A	Customer A	Product 2	2	01/09/202
Customer A	Customer A	Product 2	2	01/09/202
Customer A	Customer A	Product 7	2	01/09/202
Customer A	Customer A	Product 3	2	01/09/202
Customer A	Customer A	Product 3	2	01/09/202
Customer A	Customer A	Product 3	2	01/09/202
Customer A	Customer A	Product 3	2	01/09/202
Customer A	Customer A	Product 3	2	01/09/202
Customer A	Customer A	Product 1	2	01/09/202
Customer A	Customer A	Product 2	2	01/09/202
Customer A	Customer A	Product 3	2	01/09/202
Customer A	Customer A	Product 4	5	01/09/202
Customer A	Customer A	Product 5	2	01/09/202
Customer A	Customer A	Product 6	2	01/09/202
Customer A	Customer A	Product 7	3	01/09/202
Customer A	Customer A	Product 8	3	01/09/202
Customer A	Customer A	Product 9	2	01/09/202
Customer A	Customer A	Product 10	2	01/09/202
Customer A	Customer A	Product 11	2	01/09/202
Customer A	Customer A	Product 12	3	01/09/202

*Z.3 - **The Architecture of Speed.** Excel prefers width; Power BI prefers depth. A flat file (Left) forces the engine to scan millions of repeated text strings, killing performance and bloating file size. A Star Schema (Right) stores text exactly once. Always isolate your facts (the numbers) from your context (the descriptions) to maximize speed and scalability.*

2. The "Calculated Column" Addiction

The Why:

This is the most common technical error for new users. You need to calculate ` Profit = Sales - Cost ` . In Excel, the logical step is to add a column. In Power BI, you see the "New Column" button, and your muscle memory takes over.

Resist it. **Calculated Columns** are computed during data refresh and stored physically in RAM. They are static. **Measures**, on the other hand, are computed on the fly by the CPU only when a user interacts with a visual. They are dynamic and context-aware.

The Experience:

Think of a Calculated Column like a **tattoo**. It is permanent. Once the ink is dry, it doesn't change regardless of what clothes you wear or where you go.

A Measure is like a **tailored suit**. It changes based on the context. If you look at the measure through the lens of "2023," it fits 2023. If you look at it through "Product A," it reshapes itself for Product A.

If you try to solve every problem with tattoos (columns), you will run out of skin (RAM) very quickly.

The Truth:

You use Calculated Columns because they allow you to verify the math row-by-row in the Data View. It satisfies your need to "audit" the calculation visually, just like checking a formula in cell C2. Measures feel like "black box" magic because you can't see the result until you drag it onto a chart. Get over this fear. If you use Calculated Columns for metrics you intend to sum or average, you are bloating your file size and confusing the aggregation engine.

The Fix:

Adopt this strict rule: **Default to Measures for any numerical analysis.** Only use Calculated Columns when you strictly need to *filter* or *slice* the data by that specific new value (e.g., grouping customers into static "Gold/Silver/Bronze" brackets).

3. The "Kitchen Sink" Dashboard

The Why:

Cognitive Load Theory dictates that human working memory is severely limited. We can only process 3 to 5 chunks of information at once. When you present a dashboard with 15 visuals, you aren't providing "detail"; you are creating noise. Furthermore, every single visual on a page generates a separate query to the database. A page with 20 visuals requires 20 simultaneous queries. This causes the "spinning wheel of death."

The Truth:

Here is the uncomfortable truth about why you built that cluttered dashboard: **Insecurity.**

You didn't know exactly what the stakeholder needed, and you were afraid they would ask a question you couldn't answer. So, you dumped *everything* on the page to cover your bases. You think you're being helpful; they think you're lazy. A cluttered dashboard says, "I didn't take the time to figure out what was important, so here is all of it. You figure it out."

The Experience:

Imagine walking into a cockpit. There are hundreds of dials. That works for a pilot trained for years. Now imagine your manager—who has five minutes between meetings—walking into that same cockpit. They will panic. Your dashboard should not be a cockpit; it should be a car dashboard. Speed, fuel, engine temp. That's it. If the check engine light comes on, *then* you plug in the diagnostic computer (Drill-through) to see the details.

The Fix:

The **5-Second Rule**. If a user cannot answer the page's primary question in 5 seconds, delete visuals. Use "Drill-through" pages to hide detail until it is specifically requested.

4. Hardcoding Logic in Power Query

The Why:

Data is rarely clean. You will inevitably find misspelled country names ("USA" vs. "U.S.A.") or outdated product codes. The "Replace Values" feature in Power Query seems like a quick fix. However, this creates brittle dependencies. Hardcoding logic hides the transformation rule inside a specific step of a specific query. It is invisible to other developers and even to your future self.

The Experience:

Hardcoding is like fixing a leaky pipe inside your wall with duct tape and then painting over it. It looks fine today. But three months from now, when the water pressure changes (data updates) and the leak bursts, you will have to tear down the entire wall to find where you put that piece of tape.

The Truth:

You do this because you want to finish the report by 5:00 PM. Creating a proper mapping table feels like "extra work." But when the source data changes from "U.S.A." to "United States" and your replacement step fails silently, your report will show zero sales for the US.

You will spend hours debugging it, only to realize the error was a "quick fix" you made months ago.

The Fix:

Never hardcode business logic. If data needs grouping or correcting:

1. Fix it at the source (best).
2. Create a **Mapping Table** in Excel (e.g., Column A: "Bad Name", Column B: "Good Name"), import it, and merge it.

Make your transformations visible, auditable, and easy to update without opening the code editor.

The Learning Curve: Resources and Communities to Join

The J-Curve: Understanding Your Trajectory

Let's be brutally honest. If you walk into this thinking Power BI is just "Excel on steroids," you are going to crash. Hard.

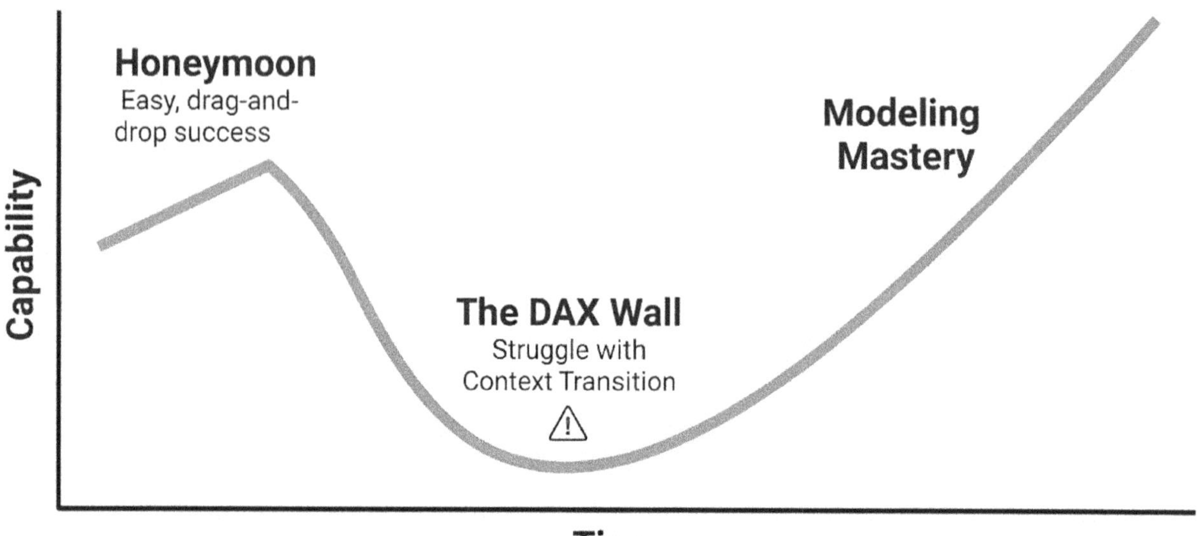

*Z.4 - **The Reality of the Learning Curve.** Everyone experiences the "Honeymoon" phase—dragging fields onto a canvas and feeling like a genius. But eventually, you hit "The DAX Wall." Your totals break, and your visuals go blank because you don't understand Filter Context yet. This trough is where 90% of Excel users quit. If you push through the discomfort and master the data model, you unlock exponential capability (Phase 3).*

There is a specific, painful collision awaiting you, known in the industry as the **J-Curve**. Here is the cognitive disconnect: Excel is *imperative*. You dictate exactly what cell A1 should do to cell B1. Power BI, however, is *declarative* and context-dependent. You define rules that shift and morph based on what the user clicks, filters, or slices. It's not just a new tool; it requires a fundamental rewiring of your mental circuitry. Most professionals experience this journey in three distinct, predictable phases:

1. **The Click-and-Drag Honeymoon.** You import a clean Excel sheet. You drag "Sales" to the canvas. Boom—a bar chart appears. You feel like a genius. This phase lasts exactly two weeks.

2. **The DAX Wall.** You need a simple calculation, like "Sales Year-over-Year." You try to type `A1 / B1`. It fails. You search Google, copy a formula you don't understand, and get a number that makes zero sense. The visuals break. The loading spinner spins. **This is the dip.** This is where 60% of learners quit.

3. **The Ascent.** You stop guessing. You sit down and learn *Filter Context*. Suddenly, the syntax clicks. You aren't just building charts anymore; you are modeling business reality.

To scale that wall without burning out, you need to ignore the noise. Focus on a curated ecosystem. Here is your survival kit.

1. The Mothership: Official Microsoft Resources

Before you spend a single dime on third-party courses, exhaust the source material. Microsoft has largely abandoned the dusty PDF manuals of the past in favor of an interactive platform.

- **Microsoft Learn (The Foundation):**
- **The Logic:** Third-party courses rot. Power BI updates monthly. MS Learn is the only source guaranteed to be current.
- **The Experience:** It's modular. You can crush a module on "Data Cleaning" while drinking your morning coffee.
- **The Reality:** It can be dry. It teaches you *how* a feature works, but rarely *why* you should use it in a messy, real-world scenario. Use it to learn the buttons, not the strategy.
- **Dashboard in a Day (DIAD):**
- **The Experience:** Free, partner-led workshops. In eight hours, you go from blank file to published dashboard. It is a firehose.
- **The Verdict:** Excellent for seeing the big picture. Terrible for retention. Treat it as a tour, not a residence.

2. The Gurus: Where the Pros Learn

When you hit the DAX Wall, the official documentation will not save you. You need deep theory and practical hacks. There are two pillars of the community you simply must know.

- **SQLBI (Marco Russo & Alberto Ferrari):**
- **The Vibe:** Imagine two Italian professors teaching quantum physics, but for data models.
- **The Why:** They literally wrote the book (*The Definitive Guide to DAX*). They don't just give you the code; they explain how the engine processes that code at the millisecond level.
- **The Reality:** They are intimidating. You might watch a video three times and still feel lost. That is normal. Their content is the medicine you *need*, not the candy you *want*. Their free "Introducing DAX" course is mandatory if you want to leave the "Excel mindset" behind.
- **Guy in a Cube (Adam Saxton & Patrick LeBlanc):**
- **The Vibe:** High-energy, fast-paced, YouTube-native.
- **The Why:** They cover the trenches. "How to make a dynamic title," "New features in the March update," or "Why is my report so slow?"
- **The Reality:** It is easy to binge-watch them for entertainment without actually learning. Treat their channel as a reference library for specific problems, not a TV show.

3. The Toolkit: Graduating from 'User' to 'Developer'

Here is a secret: Professional developers rarely build everything inside the Power BI Desktop interface.

It is too slow and clunky for serious modeling. We use **External Tools**—software that hooks directly into Power BI's engine.

- **Bravo for Power BI:**
- **The Function:** It automates the boring stuff. It instantly generates a proper Calendar Table (non-negotiable for time intelligence) and formats your messy DAX into readable code.
- **The Entry Point:** Download this first. It is the bridge between beginner and intermediate.
- **DAX Studio:**
- **The Function:** The mechanic's garage. When your report takes 10 seconds to load, DAX Studio lets you look under the hood to see exactly which calculation is clogging the engine.
- **The Reality:** It looks like a hacker's terminal. It's ugly. But it is the only way to objectively measure performance.
- **Tabular Editor (2 vs. 3):**
- **The Function:** The architect's blueprint. It allows you to batch-edit 50 measures at once or create "Calculation Groups" (a feature that stops you from writing the same measure 20 times).

4. Communities: Where to Ask for Help

You will get stuck. You will stare at an error message that says something cryptic like *"A single value for column 'Price' cannot be determined."* When that happens, you need a tribe.

- **r/PowerBI (Reddit):**
- **The Vibe:** Brutally honest, career-focused, and rapid.
- **The Strategy:** Don't just post "Help me." Post a screenshot of your data model, your code, and the expected result. The quality of your question determines the quality of the answer. This is the best place for "Is this possible?" questions.
- **Microsoft Fabric Community (Official Forums):**
- **The Vibe:** A massive, searchable archive.
- **The Strategy:** Use this for bugs. If a visual suddenly stops working after an update, this is where you verify if it's a global glitch or just you.

5. Certification: The PL-300

The **Microsoft Certified: Power BI Data Analyst Associate (Exam PL-300)** is the gold standard credential. But there is a massive misconception about what it actually means.

The Truth: A certification proves you know the definitions; it does not prove you can solve a business problem. I have interviewed "Certified" candidates who could not calculate a simple margin percentage without help.

- **The Value:** The exam forces you to learn the parts of the tool you would naturally avoid—like Row-Level Security, workspace administration, and governance. It fills your blind spots.
- **The Recommendation:** Do not take this immediately. Build three to five real-world reports first. Struggle with the data. Break a model. Fix it. *Then* study for the exam. The theory will only make sense once you have the battle scars to contextualize it.

Certification: Is the PL-300 Exam Right for You?

The **Microsoft Certified: Power BI Data Analyst Associate (Exam PL-300)**. It's a mouthful. It's also the undisputed industry standard, having retired the old DA-100 credential.

But let's get real.

You are a professional. You have quarterly closings, ad-hoc marketing requests, and barely enough time to eat lunch, let alone study. A certification needs to be more than a digital badge you slap on LinkedIn to impress your college friends. It needs to be a lever.

It must provide genuine career leverage.This exam is designed to validate a specific reality: can you handle the full "data-to-insight" pipeline? We aren't talking about making pretty charts. We are talking about scraping messy web data, cleaning it without manual intervention, and publishing a secure, mobile-optimized app for a CEO who creates data security risks just by waking up in the morning.

The Exam Profile: Anatomy of the Challenge

The PL-300 is not a memory test. If you are memorizing where the "File" menu is, you are wasting your time.

It is a simulation of an analyst's daily battlefield. The questions shift the focus from "how do I click this?" to "why on earth would I architect it this way?" It tests intent, not just navigation.

What is Actually Measured? (The "Hidden" Curriculum)

Here is the truth that catches most heavy Excel users off guard. **The exam does not care if your dashboard is aesthetically pleasing.**

It cares if your system doesn't crash. Microsoft has designed the question bank specifically to penalize "quick and dirty" Excel habits and reward scalable architecture. If your instinct is to patch a problem rather than fix the source, you will struggle here.

1. Prepare the Data (25–30%)

This tests your ability to automate cleaning. If you are still opening a CSV file to manually delete the top three header rows before importing it, **you will fail**. The exam demands you know how to use Power Query (M) to make the data clean itself upon every refresh.

- *The Trap:* Ignoring "Query Folding." You need to know which transformations happen on the server (lightning fast) and which happen on your laptop (painfully slow).

2. Model the Data (25–30%) – The "VLOOKUP Killer"

This is the hardest section for Knowledge Workers. In Excel, you merge data into one giant, wide, flat table. In Power BI, that approach destroys performance.

- *The Test:* You will face drag-and-drop questions where you must build a **Star Schema**. You need to connect dimension tables to fact tables and understand why a "Many-to-Many" relationship is usually a design failure, not a handy feature.

3. Visualize and Analyze the Data (25–30%)

It's not about choosing colors. It's about accessibility and logic. Can a screen reader interpret your chart? Have you configured the Q&A visual for AI integration? The exam tests your ability to set up drill-downs that follow a logical business hierarchy, not just what looks good on a projector.

4. Deploy and Maintain Assets (15–20%)

This is the "Solo Analyst" blind spot. If you are the only person using Power BI in your company, you likely never touch **Row-Level Security (RLS)** or **Workspace App management**. The exam assumes you are working in a rigorous enterprise environment. You cannot "wing" this section; you must study the governance rules specifically, or you will lose easiest points on the board.

The ROI: Why Bother?

For the high-efficiency professional, dropping $165 and 40+ hours on study time requires a tangible return. The value here is twofold: **External Credibility** and **Internal Maintenance**.

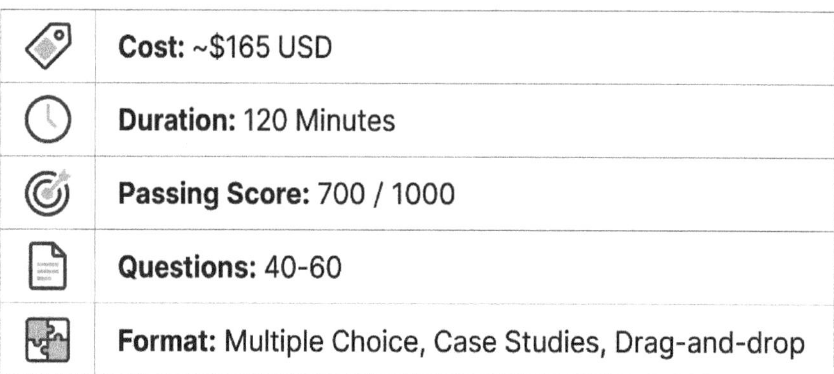

	Cost: ~$165 USD
	Duration: 120 Minutes
	Passing Score: 700 / 1000
	Questions: 40-60
	Format: Multiple Choice, Case Studies, Drag-and-drop

*Z.5 - **The PL-300 Logistical Baseline.** Knowing the rules of the game is half the battle. You do not need to be perfect; a 70% passing score (700/1000) proves your competence as a Data Analyst. Be prepared for a grueling 100 minutes of active screen simulations and dense case studies. Do not book this exam until you have built at least three end-to-end dashboards from scratch.*

The Credibility Gap

Recruiters and HR algorithms use "PL-300" as a binary filter. It serves as objective proof that separates you from the crowd of resume-padders who list "Proficient in Power BI" simply because they once opened a .pbix file. It signals to a hiring manager: *"This person knows the Microsoft standard, not just their own hacked-together method."*

The "Free Renewal" Asset

Here is the economic argument. Microsoft abandoned the old "re-test every two years" model, which was expensive and stressful.

- **The New Model:** Once you pass PL-300, you own it. To keep it active, you simply pass a free, unproctored, open-book online quiz once a year.
- **The Payoff:** The initial investment buys you a lifetime credential, provided you spend 30 minutes a year refreshing your knowledge. That is an exceptionally high long-term yield for a single effort.

Strategic Recommendation: The "3-Month Rule"

Here is the advice coaching centers won't give you: **Do not take this exam immediately.**

If you study for the exam before you have struggled with the software, you become a "Paper Tiger"—someone certified on paper but useless in a crisis. You will memorize that "Star Schema is best," but you won't understand *why* until you've tried to build a report on a flat table and watched it crash your CPU.

The Optimized Path:

1. **Phase 1: Messy Experimentation (Months 1-3):** Build real reports for your job. Break them. Google the error codes. Feel the frustration of a slow report. This creates the "synaptic hooks" that make the theory stick.

2. **Phase 2: Structured Gap Analysis (Month 4):** Now, open the official study guide. You will realize, *"Ah, so THAT is why my report was slow—I wasn't using a Star Schema."* Focus strictly on the things you can't practice at work (usually the Governance/RLS section).

3. **Phase 3: Validation (Month 5):** Take the exam to stamp "Verified" on the skills you have actually earned.

Final Words: Owning Your Value as a Data Professional

We are done with the syntax. We have covered the functions, the schemas, and the modeling. But if you close this book thinking Power BI is just about writing DAX, you have missed the point entirely. The true value of this tool isn't in the code you write; it is in the **career leverage** you create.

To close this journey, we need to have a brutally honest conversation about your market value, the economic reality of your output, and the dangerous trap of being "too good" at Excel.

From "Task-Doer" to "Asset-Builder"

Let's look at your daily grind through the lens of an economist. When you build a report in Excel—manually copying, pasting, reconciling, and praying the links don't break—you are engaging in labor that has a **1:1 correlation with time**.

Every time that report is requested, you must pay the "time tax" again. You work an hour; you produce an hour's worth of output. In this model, you are not an asset to your company. You are an expense. You are a "Task-Doer."

When you deploy a Power BI solution, the math changes. You are doing something fundamentally different. You are front-loading the pain to build a system that runs without you. The first refresh costs you ten hours of development; the hundredth refresh costs you **zero**.

Here is the shift: You are no longer paid for your time; you are paid for the *system* you built. You have moved from being a Cost Center (operating expense) to a Value Creator (capital asset builder).

The Truth About the "Excel Ceiling"

There is an uncomfortable truth about corporate life that few managers will tell you to your face: **Being the "Excel Guru" is a career trap.**

If you are the only person who understands the complex web of VLOOKUPs and VBA macros holding the department together, you have made yourself indispensable. That feels like job security. It isn't. It is a cage. You cannot move up because the moment you leave your desk, the reporting infrastructure collapses. You are chained to the spreadsheet.

Power BI breaks these chains. By documenting your logic in steps (Power Query) and clear measures (DAX), you create a transparent system that does not rely on your physical presence to function. Paradoxically, by making yourself *less* necessary for the daily grind, you become significantly *more* valuable for strategic leadership.

Market data supports this aggressive shift. While traditional Financial Analyst roles often hit a salary ceiling when the manual workload maximizes their capacity, professionals who add "Data Modeling" and "Power BI Development" to their stack break through that ceiling. The market pays for leverage, not effort.

Your New Identity: The "Analytics Translator"

You might feel a touch of Imposter Syndrome right now. You aren't a Data Scientist. You don't have a PhD in Statistics, and you probably don't write Python code.

Good. That is exactly why you are valuable.

The modern business world suffers from a massive "Disconnect Gap." On one side, you have IT and Data Scientists who understand the code but don't know what "EBITDA" or "Customer Churn" actually means for the bottom line. On the other side, you have Sales Directors and CFOs who know the business but treat data like a black box. You are the bridge. McKinsey & Company calls this role the **"Analytics Translator."**

You possess the rarest combination in the market: **Domain Expertise** (you know the business context) plus **Technical Fluency** (you can manipulate the data). A pure coder cannot spot that a Gross Margin figure "looks wrong" based on last month's promo strategy. You can.

Flipping the 80/20 Rule

There is a statistic that haunts the data industry: Analysts spend **80% of their time cleaning and prepping data** and only **20% actually analyzing it**.

Think about your last month end. How many hours did you spend fixing date formats, removing blank rows, or combining files? That is low-value friction. It is waste.

Power Query allows you to invert this ratio. Once you automate the cleaning process, you reclaim that 80%. But here is the critical part—and the challenge: **What will you do with that time?**

If you simply use the time saved to go home early, you have improved your life, but not your career. To truly own your value, you must reinvest that time into the "20%"—the analysis. You stop asking "What happened?" and start asking "Why did it happen?" and "What should we do next?"

Future-Proofing in a Microsoft World

Finally, a note on security. Technology moves fast, and learning a new tool is an investment risk. Is Power BI a fad?

Look at your office. You use Outlook. You use Teams. You live in Excel and SharePoint. Power BI is native to the very air your company breathes. It is not going anywhere. By aligning your skillset with the industry standard, you are ensuring your skills remain liquid and transferable across almost any industry or company.

The Final Call

You now have the tools. You understand that you are building assets, not just completing tasks. You are the Analytics Translator who can speak the language of the boardroom and the language of the database.

Stop settling for being the person who updates the spreadsheet. Start being the person who designs the intelligence that drives the business.

The data is waiting. Go build something permanent.

Bonus Resource Library: The Video Course & Datasets

Congratulations. You have completed the theoretical and structural foundation of Power BI. You now possess the "Architect Mindset." However, reading about a hammer is different from swinging one.

To bridge the gap between theory and execution, this book includes free access to the **Learn Power BI in 7 Days** video course and its accompanying data files. This is a practical, project-based curriculum designed to take you from a blank canvas to three fully functional, real-world dashboards.

Below is the blueprint of what you will build and the exact files you will use to build it.

The 7-Day Video Course

The video course is broken down into daily modules, slowly increasing in complexity. You will not just watch; you will build alongside the instructor.

Day 1 & 2: The Foundation (Data Cleaning & Prep)

- **The Goal:** Install the software, understand the interface, and learn how to transform "dirty" data into a clean, reporting-ready format.
- **Key Skills:** Using Power Query to rename columns, remove errors, split data, replace values, and build your very first Dimension tables (like a dedicated Country or Date table).
- **The "Aha!" Moment:** Merging a raw Fact table with a Country Dimension table to drastically improve data efficiency.

Day 3: Architecture & DAX (Modeling & Reporting)

- **The Goal:** Connect your clean tables and write your first analytical formulas.
- **Key Skills:** Understanding the difference between Star, Snowflake, and Galaxy schemas. Writing DAX to calculate KPIs like *Average Response Time* and *Average Satisfaction Score*.
- **The "Aha!" Moment:** Dragging your new DAX measures onto the canvas to instantly create interactive Slicers, KPI Cards, and Bar/Line charts.

Day 4 & 5: Advanced Logic & AI Bonus (Sales Data)

- **The Goal:** Tackle complex business logic, like calculating revenue with dynamic promotions.
- **Key Skills:** Using advanced Power Query transformations (Custom and Conditional columns) and correcting DAX errors. Building a comprehensive Sales Report with Product, Store, and Campaign slicers.

- **The "Aha!" Moment:** The **AI for DAX Bonus Module**. You will learn how to write clear prompts for ChatGPT to generate complex DAX code, and more importantly, how to validate that AI-generated code to ensure it is mathematically accurate.

Day 6 & 7: The Executive View (Project Construction)

- **The Goal:** Build a financial monitoring dashboard and share it with the world.
- **Key Skills:** Calculating advanced variances (Budget vs. Actuals) and tracking *Completed vs. In-Progress* tasks. Designing a single-page management dashboard.
- **The "Aha!" Moment:** Publishing your `.pbix` file to the Power BI Service, configuring scheduled refreshes so the data updates automatically, and reviewing essential tips for passing the PL-300 Microsoft Certification exam.

The Dataset Vault: Your Practice Files

To complete the 7-Day course, you have been provided with a vault of raw data files and completed reference models. Each dataset is designed to teach a specific business scenario.

Dataset 1: Customer Support (`customer_support.xlsx`)

- **When to use it:** Days 1 to 3.
- **The Scenario:** You are managing a global IT helpdesk.
- **Key Columns:** `TicketID`, `Priority`, `Status`, `CreatedDate`, `ResponseTime`.
- **The Lesson:** This file is your introduction to data cleaning. You will learn how to handle null values in response times and build visuals showing how many tickets are currently open per country.

Dataset 2: Retail Sales (`Sales_Data.xls`)

- **When to use it:** Days 4 to 5.
- **The Scenario:** You are analyzing revenue for a retail chain running complex promotions.
- **The Lesson:** This file forces you to master DAX. While it contains standard quantity and price columns, the real challenge is the `Promotion` column. You must write formulas that dynamically calculate the final discounted price based on offers like "20% Off" or "Buy One Get One Free" (BOGOF).

Dataset 3: Financial Monitoring (`Project_Construction_Data.xlsx`)

- **When to use it:** Days 6 to 7.
- **The Scenario:** You are tracking the budget and progress of massive construction projects.
- **Key Columns:** `Task Name`, `Phase`, `Budget`, `Actual`, `% Complete`.
- **The Lesson:** This is all about financial variance. You will learn how to calculate the difference between forecasted budgets and actual spend, and visualize the completion status of distinct project phases (e.g., Foundation vs. Site Work).

- **When to use it:** Day 1 (and ongoing practice).
- **The Scenario:** A classic relational database featuring `Orders`, `Products`, `Customers`, and `Employees`.
- **The Lesson:** We introduced this file in Chapter 1 for your very first "Quick Win" dashboard. Beyond that immediate victory, it serves as the ultimate test of your Data Modeling skills. The data is deliberately split across multiple files/tables. You must correctly link Customers to Orders using the `CustomerID` to build a functional Star Schema.

The "Cheat Codes": Reference Files & Assessments

Do not get stuck. If your dashboard looks wrong or your DAX is throwing an error, use the provided solution files to reverse-engineer the correct answer.

- **`NorthWind_Sales_Report.pbix`:** Open this to see exactly how a professional architects the relationships between the Northwind tables and applies corporate design colors.
- **`Project Report.pbix`:** The finished model from Day 7. Use this to study how Slicers are configured to make the financial report perfectly interactive.
- **The PL-300 Prep Test (`PowerBI_Basics_MCQ_Questions & Solutions`):** Two Word documents containing a 30-question Multiple Choice test. Use this to validate your theoretical knowledge before you start building, or as a diagnostic tool before attempting the official Microsoft exam.

How to Access Your Materials

Please note that every single dataset, reference `.pbix` file, and assessment listed in this appendix is an **integral part of the video course**. You do not need to hunt for standalone download links or navigate a messy cloud drive.

To access these materials, simply scan the **"WATCH IT IN ACTION"** QR codes conveniently located at the end of each chapter (Days 1 through 7). Those codes grant you direct access to the daily video modules, where all the specific files required for that day's lesson are hosted and ready for immediate download.

About the Authors

Kora Pierce & Callum Pierce

The transition from a manual spreadsheet user to a modern data architect requires two distinct types of learning: understanding the *strategy* of how data works, and mastering the *execution* of the software. That is why this hybrid learning experience was built by a specialized team combining both perspectives.

As siblings and analytics professionals, Kora and Callum spent years watching highly intelligent people waste their Fridays wrestling with broken VLOOKUPs, manual exports, and freezing files. They realized they were tackling the exact same corporate problem from two different angles. They combined their strengths to create a complete, friction-free learning system.

Kora Pierce

Creator & Instructor, *Learn Power BI in 7 Days* Video Course

Kora is the voice and screen-presence behind the video curriculum. After years in the corporate reporting trenches, she grew frustrated with the gap between knowing what the data *should* say and fighting the software to actually show it. Kora specializes in translating intimidating software interfaces into intuitive, step-by-step workflows.

Her teaching philosophy is ruthlessly practical: show the exact clicks, explain the immediate business value, and skip the unnecessary jargon. She designed the video course to be the ultimate "over-the-shoulder" experience, ensuring that learners can watch the interface come to life in real-time. When she isn't recording tutorials or designing executive dashboards, Kora is constantly hunting for the absolute fastest path from raw data to a finished, beautiful report.

Callum Pierce

Author, Power BI for Excel Users

Callum is the primary author and the structural mind behind the book. With a deep background in dismantling fragile, overly complex reporting systems, Callum is a relentless advocate for the "Architect Mindset." He has spent his career helping organizations transition from reactive "Data Janitors" to proactive data strategists.

Callum's expertise lies in data modeling, DAX logic, and system governance—the invisible architecture that makes a dashboard actually work at scale. He wrote this book to provide the theoretical anchor to Kora's practical execution. His goal is to ensure that learners don't just memorize where the buttons are, but truly understand the physics of their data, allowing them to build reporting engines that run on autopilot.

www.ingramcontent.com/pod-product-compliance
Ingram Content Group UK Ltd.
Pitfield, Milton Keynes, MK11 3LW, UK
UKHW061706190726
13853UKWH00008B/2442

9 798904 177034